I0610052

Rowland Ward

The English angler in Florida

With some descriptive notes of the game animals and birds

Rowland Ward

The English angler in Florida
With some descriptive notes of the game animals and birds

ISBN/EAN: 9783337238032

Printed in Europe, USA, Canada, Australia, Japan

Cover: Foto ©berggeist007 / pixelio.de

More available books at **www.hansebooks.com**

THE

ENGLISH ANGLER

IN

FLORIDA

WITH SOME DESCRIPTIVE NOTES OF THE GAME
ANIMALS AND BIRDS

BY

ROWLAND WARD, F.Z.S.

AUTHOR OF 'RECORDS OF BIG GAME,' 'SPORTSMAN'S HANDBOOK,' ETC.

WITH NUMEROUS ILLUSTRATIONS

LONDON

ROWLAND WARD, LIMITED

166 PICCADILLY

1898

PREFACE

TARPON fishing sport was for many years a kind of mystery to both American and English fishermen, but within the last decade it has become well known, and has even developed into something like a "boom." At least fifteen years since a gentleman who had caught several tarpon on rod and line, calling upon me in Piccadilly, showed me one of the scales he had brought home from Florida; and a striking object it was, from its immense size, its horny substance, and, most of all, from that partial covering of fine whitish silver which, at a casual glance, looks as if it could be dusted off with the light flick of a handkerchief. From that time forward I kept the tarpon in view as an honest object of desire, and all that I read about it, and the specimens which came under my notice as a taxidermist, confirmed my determination to kill one for myself some day.

Last spring Mrs. Ward and myself accordingly sailed in the

s.s. *Majestic* on the 24th March.   We were home again by the end of May.

Meanwhile, some account of the sport I enjoyed had been published in the *Field* and other papers, and thereafter I was often asked to put together my notes and publish them in handbook form, for the guidance and assistance of brother anglers who might meditate a visit to Florida.   As the camera is always an inseparable companion for me on such expeditions, I brought back a large collection of views.   Owing to unfriendly weather these were often much more imperfect than I should have wished, but they are at least genuine. They would have been better in some cases if I had taken a stand and could have used it for time exposures.   What I have are certainly better than nothing—infinitely better than the vamped-up illustrations that are often given to the public. The requests made to me, and the possession of these original photographs, have at length induced me to make the attempt (from which I confess I at first shrank) of recording my experiences in Florida ; and I do it, as an enthusiastic sportsman of long standing with rod as well as gun, and also as something of a traveller, in the hope that what I have to say will be useful to many.

CONTENTS

The Book of the Tarpon, by A. W. Dimock, with ninety-three photographs by J. A. Dimock. (Frank Palmer.)

DURING recent years several works dealing with tarpon-fishing have appeared in print. When first anglers read that excellent book "Tarpon Fishing in Mexico and Florida," by Mr. E. G. S. Churchill, they wondered how it was possible for any photographer to get such snap-shots of fish leaping as were then reproduced. We have now from the pen of Mr. A. W. Dimock, an American angler, a book which graphically describes how and where to catch tarpon, and also how to hold them so that the camera-man may get his pictures. In this case the man behind the camera was Mr. Julian A. Dimock, and in the ninety odd photographs reproduced he has given us the best series of tarpon-fishing pictures yet seen. Some of the illustrations are wonderful, and every incident in a day's tarpon-fishing, from the moment of hooking a fish until it is finally brought to the gaff, is clearly shown by the camera. Another well-known American author, Mr. C. F. Holder, has described fishing for the "silver king" as being "without doubt the most sensational and exciting of sports, and one of the most dangerous when persistently followed." These remarks may well be applied to tarpon-fishing if carried out in the method employed by Mr. Dimock and his companion, for they fished entirely from a small Peterboro' canoe, and often using only a light trout rod. Most anglers find that playing tarpon from a good-sized boat with a strong rod is sufficiently sensational ; but Mr. Dimock seems to relish the additional excitement of often having his small craft capsized by the mad rushes of a big fish. The author writes modestly of his own exploits and of the risks which he takes in landing these game fighting fish under most difficult conditions. To give an idea of what sport he has had we may quote the following extract from his book. During two months' fishing on the coast of Florida in fifty-two days he captured 334 tarpon, and "the tarpon varied from 1½lbs. to a hundred times that weight, and in length from 18in. to 6½ft. All were taken from a light Peterboro' canoe, and sixty-three of them on an eight-ounce fly-rod." This, indeed, must have been something like sport, and we thoroughly recommend anyone who is contemplating a tarpon-fishing expedition to get a copy of this most interesting book.

24.8.'12

TARPON PROBLEMS.—In your issue of June 29 your correspondent "F. G. A." propounds certain questions concerning tuna and tarpon. I am unable to give him any information concerning tuna, never having caught them, but I can perhaps answer some of his questions about tarpon. In addition to his question, "What is the maximum weight of the tarpon?" I would state that the largest tarpon ever caught at Aransas Pass, Texas, U.S.A., measured 7ft. 6in. in length, and weighed 225lb. This fish is mounted and set up in the office of the Bayside Inn, at Rockport, Texas. As to the probability of there being larger tarpon, I can only state that during the last eleven years at Aransas Pass, probably 7000 tarpon have been caught and measured, and this fish is the largest of them all. Your correspondent's second question concerning the food of the tarpon, would seem to imply that he uses cut bait and fishes in 45ft. depth of water. The bait used at Aransas Pass is the natural food of the tarpon, viz., a whole mullet about 6 to 8in. long. This bait is trolled very near the surface of the water, and, in fact, is frequently cast in the path of the tarpon, whose approach can often be plainly seen. It may be of interest to describe the tackle prescribed by the Tarpon Light Tackle Club of Port Aransas. The rod (composed of a butt and a tip), when assembled, must not be less than 6ft. in length. The tip must not be less than 4ft. 6in. in length, and must weigh not more than 6oz. The butt must not be over 18in. in length. The linen line must not be larger than a nine-thread line. It would be interesting to learn how the above-described tackle compares with that used when fishing for tarpon in 45ft. depth of water with cut bait.—H. H. C.

6.7.12

ILLUSTRATIONS

THE ENGLISH ANGLER IN FLORIDA

I

FROM ENGLAND TO FLORIDA

THE voyage from Liverpool to New York is, by this time of day, too well known to require elaborate description. The track is a well-beaten ocean highway, and a run across "the herring pond" is seldom of little interest to other than the passenger. Our experience, however, suggests that I ought to remind the reader that if he sails after the middle of March he may expect equinoctial gales. We had them, and obstinate head seas, until the 30th, when we congratulated one another upon our first really fine day; and the assurances that had been given us of arrival on the 1st of April were, as a matter of fact, duly honoured by our bringing up in the river about

B

five o'clock in the morning to await the pleasure of the medical examination.

Unless the traveller to Florida has brought his sporting or other equipment from the other side, he will find it no **New York purchases.** loss of time to remain a few hours in New York, and more especially if he has to provide himself with the fishing tackle necessary for tarpon and other fish. The way, as to ship and train travelling, had been made delightfully easy to me before starting. I went to Thomas Cook and Son, placed myself in their hands, and after calling at their office in New York there was no further trouble. We and our baggage were straightaway booked to the end of our land journey, namely, Punta Gorda, West Florida. Roughly speaking, London is nine days from this place.

Before purchasing fishing tackle we went off to the Museum, where I wanted to examine the casts of Florida fishes, but as they were very indifferently done, and not named, I did not waste much attention upon them, devoting what brief leisure I had to the stuffed alligators, birds, and other animals; the birds well represented with their natural surroundings. Here also I was deeply interested to find a picture of Audubon,—with whom my father travelled over fifty

years ago,—and the gun he used.   By the advice of Mr. J. A.
Jameson, who had special knowledge of the subject, I bought my
tackle at E. Vom Hofe's, a practical tarpon fisherman himself,
and a well-known frequenter of the Florida waters.   What I
bought will be described presently under the heading "Tackle."
It is mentioned here as a reminder, because although tackle
may be procured in Florida, New York is the best place
for the purpose.   A good map of Florida should not be at
this preliminary stage forgotten ; there will be ample opportunities of studying it *en route.*   The best map I could procure
was the Indexed County and Township Pocket - Map and
Shipper's Guide of Florida, published by Rand, M'Nally, and
Company, of Chicago and New York.   It is well indexed,
and the rivers are clearly indicated.

The railway trains are of the customary American type—
Pullman sleepers, and drawing-room cars for those   **Journey**
who prefer that isolated luxury.   Passing as you   **South.**
do by Philadelphia, Baltimore, and Washington, and speeding
southward through Virginia and the Carolinas, past Charleston
to Georgia, under any circumstances the run is replete with
interest to the English traveller.   Although our day of start
was the 3rd April, we found it intensely hot in the train.

There was a buffet on board with the usual coloured attendants ;
but the sight of tinned goods might lead some persons to

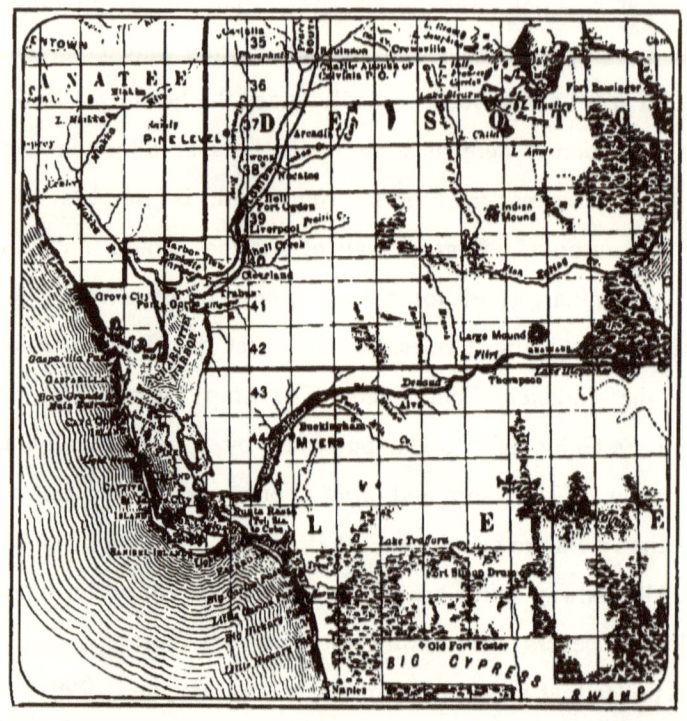

FROM THE INDEXED COUNTY AND TOWNSHIP POCKET-MAP AND SHIPPER'S GUIDE.
Published by Rand, M'Nally, and Co., of Chicago and New York.

consider whether it would not have been more agreeable to
have arranged a portable commissariat, along with the other
stores, in New York.

It may be here mentioned that you may either travel through direct to Punta Gorda, sleeping two nights in the train, or break the journey at Jacksonville for one night. We elected to follow the latter course, and left at 9.30 A.M.; the other train started somewhere about five o'clock P.M.

There is yet another route, namely, the sea voyage from New York to Jacksonville. The steamers are large and luxuriously appointed for passengers, and run frequently and regularly during the season. This is indeed the favourite route for the New York people, who are habitual visitors to Florida. It is the cheapest mode of reaching Jacksonville, and if a little longer in point of time, has the compensating pleasures of life on board ship.

We were too early in the season for green foliage and spring flowers in the northern half of the journey, and, besides, there were rains, clouds, and high winds. All day on Sunday (4th April) we were rattling through dreary Georgian pine forest and swamp, with occasional glimpses of wild-bird life, notably flocks of gulls, the species of which we were unable to identify. The weatherboard houses that came into view looked very desolate under these circumstances. One of the not un-common incidents of long-journey travel in the States kept us fretting at a long delay within half an hour of the time at which we were due at Jacksonville.   One of the swamp bridges was on fire, and the repairs kept us six hours waiting.   Some of the car darkies, who seemed to be frightened out of their timid lives, were afraid of " crackers " (as the blacks of solitary habits inhabiting those wilds are called), and it was evident that you had only to put on something white, and play the ghost, to scare them into fits.

Jacksonville is but a few miles south of the river boundary of Georgia, and was our gateway to Florida.   It is a really

Entering important city, and till within comparatively
Florida. recent times a reference to Florida, even to the untravelled American, meant Jacksonville and little else.   Railway lines now centre there from all points of the compass, and it has its Government buildings and a choice of fine hotels.   But it was night when we reached the station, and we were not in the humour to admire or get up any enthusiasm, with the heavy rain and humid atmosphere in which we found ourselves.   The comforts of the great Windsor Hotel, where a few smart people were still staying, were, however, cordially appreciated.   We found a good dinner ; and at 8.30 next morning, after early bath and breakfast,

resumed our journey. The season was now over, and the hotel was closing in a few days.

Now in truth we were soon able to understand something of the real Florida, and the enthusiasm with which it is regarded as a health resort in the Northern States, whence hundreds of citizens migrate from December to April, to find warmth and beauty in place of merciless frosts and snows.   There are two aspects of Florida: (1) the dismal swamps and wildernesses haunted by alligators, crocodiles, pumas, and rattlesnakes ; (2) the fragrant orange groves, palm-trees, and semi-tropical shrubs and flowers, bright home of gorgeous butterflies and birds. Beautiful, in the lovely morning, after the gloom of the previous day, were the orange plantations, and the graceful fan palms towering above the forest trees.   From some of the latter, festoons of moss-like growths hung like banners.   All around, the greenery was as that of an English June.   Here were the pine and other trees tapped to furnish materials for the turpentine industry.   Flights of buzzards were in the air. The rivers we crossed, with their banks lined with palms, would have reminded me of the Upper Nile but that the ground was bright in its emerald herbage.   One pool we passed was a reservoir of flowering water-lilies, and there were butterflies on the wing of the deepest orange colour. This was indeed typical Florida, and it was not difficult to comprehend why the visitors from the New England winter hail it as their Italy, their " own Riviera."

However, before the eighteen hours in the train were at an end, we longed for a change. It was in truth at the last part rather a tedious journey. The principal meal of the day was taken during a halt at Lakeland, where we changed to a branch line, the express continuing westerly to Tampa, called " the Magic City of the Gulf " ; it is the fashionable resort of pleasure-seekers, and a rival to Jacksonville for the metropolitan honours of the State. It was slow travelling from Lakeland, and it was half an hour after midnight when we reached our terminus, the town of Punta Gorda.

THE RAVALLIA.

II

ON THE FISHING GROUNDS

IT was something to be thankful for that we were at last at Punta Gorda, and actually on the fishing grounds of Florida, on the coast where tarpon abounded, and where **Punta** the waters teemed with other game fishes, not per- **Gorda.** haps so distinguished for heavy weights, but beloved by the angler. Had not one of the pamphlets lying about the saloons in the train informed us? "Here indeed is the ideal resort of sportsmen and anglers. Located just below the 27th parallel of latitude, the temperature never goes below 40 degrees in winter, and frost is unknown." There was also an alluring paragraph about a modern first-class hotel beautifully located in the bay; and at half-past twelve at night that was quite as important as, "the scenery is all that appeals to the sentiments of a lover of Nature." But we had literally reckoned without our host. The season here was over, and that modern first-class hotel was shut up.

The dilemma, however, was not very serious, as we intended to go on to Fort Myers by the first steamer, and by

ON THE FISHING GROUNDS.

good fortune we found that there would be one starting in the morning, and that we might go on board and take our berths. This we did there and then, obtaining, with the welcome accommodation, our first acquaintance with the mosquitoes of Florida, which did us the honour more or less to keep us company during the whole of our stay in those parts.

Although some days were to pass before I caught my first tarpon, continual references would be made to that fish

TWO TARPON.

by the sportsmen seen or heard of on our sailings and excursions in this interesting fishing country. As a rule, people there think and talk of nothing else, and I was much amused one day when a gentleman estimated the proportions of a lady about whom we were conversing by the remark : " Well, she is a very fine woman ; weighs quite as much, I should say, as a 160 lb. tarpon."

Being now, therefore, at Punta Gorda, on the scene of action, so to speak, I will take the opportunity of unburdening myself of what I know about the tarpon and how to catch it. Also, this is the convenient time for informing the reader that Mrs. Ward undertook the task of making such rough notes of our daily doings as we thought worth preserving ; and when the tarpon, in its general character, has been disposed of in the following pages, the entries from her Diary will be added for such particulars of sport and travel as she recorded.

The tarpon bears a variety of names, some of which are already falling into desuetude, though occasionally it is still **The Fish** spelt tarpum. He is " The Silver King " and **itself.** " The Big-Eyed Herring " of his admirers. The late Professor Brown Goode deals with it in the chapter in his work entitled " The Herring and its Allies." He mentions it in connection with the shads, and says the most important member of the family is the Tarpum, or *Megalopus thrissoides*, or *M. atlanticus.* At the time the work was written not a

great deal was known of the fish as providing grand sport with
the rod and line. Indeed, some notes contributed by Mr.
Stearns may be interesting to quote upon the point, as, in
some respects, showing the somewhat imperfect knowledge
which existed upon the subject within comparatively recent
times :—

The Silver Fish, or Grande Ecaille, is common everywhere on the
Gulf coast. It is an immense and active fish, preying eagerly upon
schools of young fry, or any small fish that it is able to receive into its
mouth, and in pursuit of which it ascends fresh-water rivers quite a long
distance. During September 1879 I saw large numbers of Silver fish
eight or ten miles up the Apalachicola River, and am told that that was
not an unusual occurrence. They go up the Homosassa River in
Florida, and several of the Texas rivers, so I have subsequently learned.
The Tarpum will take a baited hook, but it is difficult to handle, and
seldom landed. The Pensacola seine fishermen dread it while dragging
their seines, for they have known of persons having been killed or
severely injured by its leaping against them from the seine in which it
was enclosed. Even when it does not jump over the cork line of a
seine, it is quite likely to break through the netting before landed. I
have secured several specimens, the smallest of which weighed 30 lbs.,
and the largest about 75 lbs. The Tarpum is said to be palatable and
well flavoured.

Professor Goode adds that the sailors' name for the fish at
Key West, Bermuda, Georgia, and elsewhere, was tarpum or
tarpon, while in Florida it was commonly called Jew fish.

The exact range of this fish is really unknown, but I have heard of some being caught at one portion of the coast of

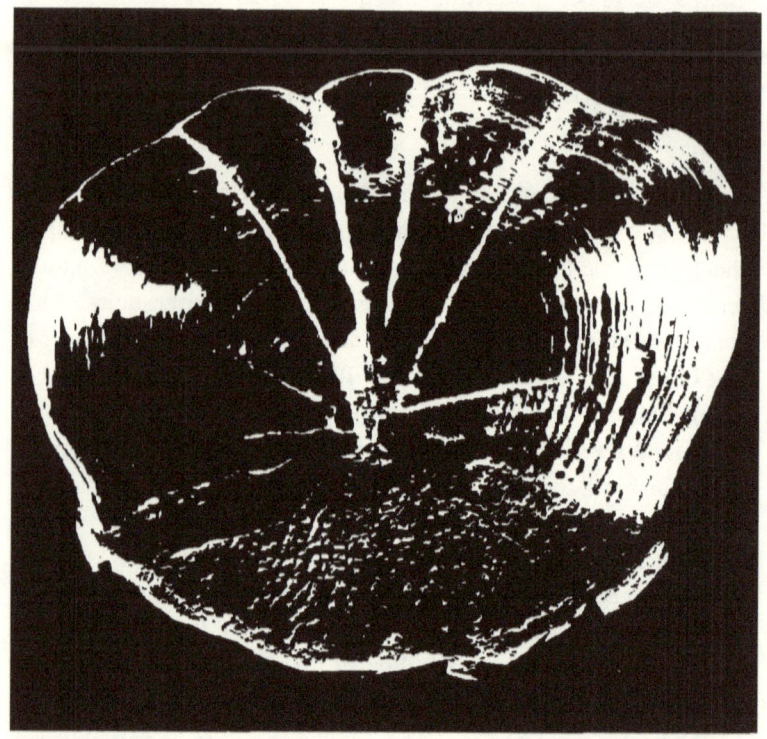

TARPON SCALE—ACTUAL SIZE.

Jamaica, and it has occurred in other portions of the West Indies, and even the Bermudas. This gigantic herring is by this time not altogether a stranger to the British public, as large specimens, very well set up, have been exhibited from time to time, and they will soon be in every museum.

The large scales, which are from 2 to nearly 4 inches in diameter, with about a third of the upper curve covered with beautiful silvering, are also not unknown, for many successful fishers for tarpon, although they do not bring home the fish itself, preserve the scales, which go very easily through the post in a cardboard box, and are always acceptable as a trophy.

III

SPORT WITH TARPON

IT must not be supposed that tarpon are caught every day; indeed you have sometimes to angle in vain for several days without a strike. They are, in fact, like all other fish in the world, capricious, and you have to exercise the virtue of patience, and be thankful when you get them in the humour.

The tackle used for this big strong fish should be of the

Tarpon Tackle. very best; hence my advice to the English angler proceeding to Florida is to expend a portion of his time on landing in New York so as to put together his equipment thoroughly.

The rod is generally a bamboo pole, as shown in the

The Rod. frontispiece illustration, with a short butt of about $1\frac{1}{2}$ feet in length, and the entire length of the implement should not be more than 7 feet. After the more or less limber salmon and trout rods of the old country, this rod handles very strangely at first, but you will soon fall into the right manner of using it ; it proves very powerful, and it is surprising how far you can cast with it.   A tarpon rod can be bought in New York for £2 or less, and those at the figure I name, as I can vouch from experience, are really good. As there is always a chance of killing a 200 lb. tarpon, or something heavier still ; as, moreover, there are accidents by flood and field to be always guarded against by the prudent sportsman, I should advise taking three rods and sets of tackle, although I only took two.   Mr. E. Vom Hofe, of New York, is, as already stated, himself a very successful tarpon fisher ; he has therefore taken great pains to perfect all details of tackle for tarpon fishing after consultations with brother anglers, and, from his and their experience, is therefore both practical and reliable.

The best class of butt, including German silver reel fittings, costs about four dollars, and the six-foot tips are made of either greenheart, split cane, or plain bamboo, and range between five and twenty dollars each.   Most of the habitual tarpon fishers have rings on opposite sides of the rod, and the end ring lined with a movable collar of agate.

No doubt the rush of a tarpon, with 200 yards of line on the reel, if very often repeated, would soon make short work of any ordinary soft metal, but I never suffered at all from not having

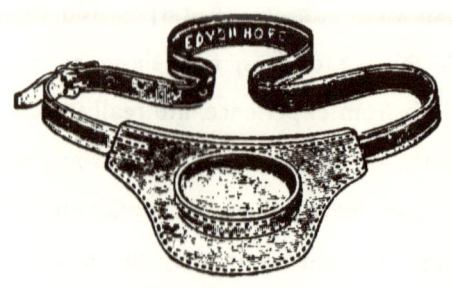

THE FRIENDLY SOCKET.

an agate ring. The rod I found most in use during my last visit was the one illustrated, with the butt whipped round with thin cane as with some of our English trout rods.

It is not customary to have a large knob at the end of the butt, and this must be considered when it is remembered

The Friendly Socket. that at times an enormous strain is laid upon the personal strength of the angler. The first tarpon I caught, for example, took me one hour and a half to kill in rough water, holding on by sheer force all the time. Hence the use of the friendly socket, which is fixed to a leather belt. It is of course obtainable at the tackle store. This simple contrivance is a real convenience, and should not therefore be omitted in the equipment. Some anglers use thumb-stalls, either thickly knitted or of leather, to protect the fingers. At one time they were considered

Tarpon line (actual size).

The check or drag.

TARPON REEL.

indispensable for manipulating the somewhat fine line when it is being rushed off the reel ; but I preferred the leather stall or drag which is fixed to the reel ; and I noticed that the majority of tarpon men from the States who go to Florida season after season rarely use the thumb-stall.

The reel, as illustrated on p. 19, is the item of tackle which should be most carefully selected.   It often occurs that the battle is lost or won by it.   My purchase was **The Reel.** E. M. Vom Hofe's striped bass or tarpon reel, described in his catalogue as "finest quality rubber and German silver, full steel pivot with German silver bands, S-shaped balance handle to screw off, sliding oil cap, tension click drag." It is a beautiful piece of mechanism, runs on ball bearings, but it is, from an English point of view, very expensive.   It is by the attached leather drag or guard, as seen in the illustration, that you are enabled to put on the necessary check when the line is running out rapidly.   As you suffer very much from the absence of this, you will never, after paying the penalty of forgetting it, go without it.   Cautious use is required if you would avoid cut or broken fingers.   The reel should hold from 150 to 200 yards of line, for it is always wise in this, as in other fishing, to be in a position to cope on equal terms with the very largest fish.   I used my Vom Hofe reels pretty hard, and it is but fair to say that they never played me tricks, or failed in the slightest degree at a pinch.   The novice will learn by experience how to avoid getting foul of the handle— an accident that has been known to result in a ̄ broken finger. Tarpon reels vary in price, and they are made of a hard composition that renders the salt water harmless.   Of course, with the work they have to do, they are on the multiplying principle. The best tarpon reel in the market, at the time of my visit, so far as I saw, was that patented by Mr. Vom Hofe, and. it cost about £7.   I had two sets of tarpon tackle (besides a general set), but most people take three.

The English stranger never fails at first to express his astonishment at the comparative fineness of the line, and its

The Line.   seeming soft texture.   The Americans make very good linen lines, however, and that for tarpon is excellent.   Even the smallest-sized of the above-mentioned reels is made to carry 200 yards of the thread linen line.   Lightness is therefore the primary essential aimed at, and however much at first you may be incredulous, you will at length admit that the end justifies the means.   At the same time, my opinion is that these lines would be better if they were soaked in stearine, or some such preparation that would keep the brine out, and would lessen the trouble of looking after when brought back to England.   Mr. Alfred Harmsworth, in the chapter he contributed to the Badminton volume on Sea Fishing, makes the suggestion that tarpon lines should have a distinct colouring for each 50 yards or so ; his reason is that when the line gets wet and swells, it is difficult to tell how much is left on the reel.   Silk lines have been tried, and other materials, for fishing in Florida, but the most experienced of the tarpon men stuck loyally to the thread line which I have described, and which is figured on p. 19.

What in England we call a trace, collar, or snood, is in America called a snell, and this is probably, next to the winch, the most important part of the tarpon

The Snell.

equipment.   The type which I adopted and adhered to was made of a strip of raw hide, and that is what is figured in the illustration in front of tarpon.   The principal objection to this leather snell (though at the same time it is an advantage) is the ease with which it is bitten through by the sharks which you are continually hooking on the tarpon grounds. If you would avoid this vexation, and can put up with considerable loss of time, you may use a snell of cotton cord, loosely woven, and wound around with copper wire. (The snell that is armed with wire is often woven so as to appear something like the wick of a lamp.) I maintain, however, that, all things considered, the raw hide snell is best, for when your bait is taken by a shark, there is an immediate severance, and an end of the trouble. This strip of leather is fastened to the line by a loop.

There is a regulation hook made for tarpon, and in the illustration it has been photographed its full size.

The Hook. The reader may remember that Mr. Stearns, quoted by Goode, refers to the difficulty, experienced at the date of his writing, in hooking the tarpon. The truth is, the interior of the tarpon's mouth is most curiously plated, and there are few vulnerable spaces into which the barb of a hook can penetrate.

There are holding places somewhere in the roof and around the rims of the "lips," and once I hooked a tarpon in the tongue. For the rest, a hook has as much chance of penetrating the sides of a tarpon's mouth as of a stone wall.

THE TARPON HOOK.

The hook that will pierce that plating has not yet been fashioned. What is used for tarpon has therefore been designed for service with gorge bait; its destination must be the gullet of the fish, and once there, there is not much chance of a breakage or ejectment, if the hook be of the right length, substance, pattern, and temper. Such a tarpon hook as that illustrated seemed to me to ably fulfil each of these requirements.

For the one pattern of hook used by the tarpon fisher there is generally one kind of bait, namely, a piece of the common gray sea mullet. There are seldom times or places on the coast when an ample supply of bait

The Bait.

cannot be secured, since the elevation of tarpon catching to a popular sport has resulted in the creation of the businesses of guides (or gillies), bait-catchers, and hangers-on. The mullet is cut into three parts, and the tail section is generally chosen for bait. All scales are removed, but the head is preferred by some fishermen, and I think Mr. W. Ashby Jones caught his big six feet nine inch fish (172 lbs.) with this. He told me that in his opinion the tail was a " mean bait." The hook is passed so that the snell is threaded through the lump of fish, which will be per- haps five inches in length, and therefore heavy enough to be cast out without a leaden sinker.   The narrow part of the bait is upwards, and the barb of the hook sufficiently embedded not to assert itself too rudely during gorging, yet not too far to admit of ejection should the tarpon feel that way inclined.

There is no fishing for tarpon from " the bank." You have to go afloat, cruise hither and thither in pursuit of the **The Angler** shoals of fish as they rove in search of food over **in action.** the oyster beds, amongst the marine vegetation, or on the shallows of the lagoons. To proceed with comfort, you hire a sailing sloop or steam launch, taking in tow the flat-bottomed, square-sterned fishing-boat which is in general use on the waters of the Floridan coast.   It has a revolving arm-chair fitted to the after thwart.   I saw a very nicely arranged keel boat with centre-board, however.   It belonged to Mr. R. T. Holloway, who used only this kind of boat, and got about remarkably well in the river.   You often have a sail of twenty miles before finding your fishing-ground, and all the while your glasses will be sweeping the surface of the water on the search for signs and tokens of fish movement.

At last you are in sight of a shoal of tarpon.   They may be in commotion, either breaking the water sharply, showing their backs in the slow and regular movements of the porpoise,

or swimming with the dorsal fin, and its queer whip-like
terminal, hoisted above the surface.   The anchor is dropped,
sails are taken in, and if the ground is suitable for gorge bait

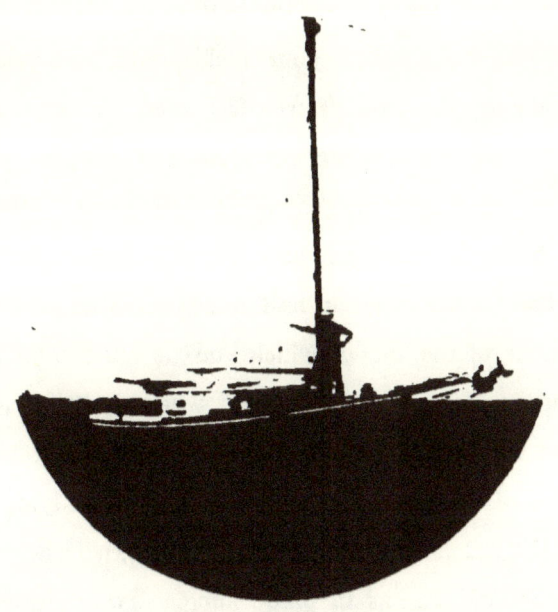

SLOOP HIRED FROM MYERS FOR TARPON FISHING ON THE RIVER.

operations the angler and his man get into the rowing-boat,
and pull say 100 yards from the yacht, though sometimes the
pull is a matter of a mile instead of yards.   The small boat is
anchored, and sometimes a pole is fastened near the stern to
keep it steady.   Your guide, indeed, will take his bearings and

choose his ground just as our Thames fishermen select their swims. It may be that you have the great tarpon all around you, and can see their swirls as the fish, head downwards, are foraging amongst weeds or on the bottom.

Now you make ready your tackle, and bait your hook. The reel already described is well fashioned for the process of

At Anchor.

casting Nottingham style, and a 20 or 30 yards cast or more should be aimed at. When the tarpon is biting heartily he is very free and fearless, and I have had my bait snapped within half a dozen yards of the boat. The portions of the mullet which you do not use for bait are broken up and cast in for "chum," which is the local name for ground bait. It often happens that this is the prelude to a long period of waiting. There yonder is the bait lying on the bottom ; the rod is placed down in the boat, with a few yards of line coiled off free, as in pike-fishing. Two rods, and even three, are sometimes used, one from each bow and one from the quarter. Your man will handle those in the bows, and it is your business to sit on the thwart facing the stern and look after your own affairs.

Sometimes for hours you are worried by the dirty, slimy cat-fish, which simply swarm on these grounds. I believe there are two or three varieties of them, but the grunting ugly beggars are a terrible nuisance, very frequently spoiling your baits one after another. You may get a sudden   **Some** rush with a shark, which had better be off than **Nuisances.** on, or you may be bothered by crabs which fool about the bait most aggravatingly, and manage at last to cut the line, you being all unconscious at the time of what has happened.

By and by you will see the line running out a few yards and stopping, and that should be your warning that a tarpon has seen and taken your offering of dead mullet. **Tarpon at** The fish may keep on running in greater or smaller   **last.** spurts, but anyhow your business is to let him gorge the bait thoroughly. The line must not be checked for a moment ; so slight a hindrance as the obstruction of a weed or blade of submarine grass coming in contact with the line has been known to result in the tarpon rejecting the bait.

By this time you have your rod in hand, ready for events, and ever watchful that the line is free. Take time ; let the tarpon gorge at his ease. In this, as in some other kinds of fishing, there is more danger in being too hasty than in being too slow. At the same time there is a limit to caution, and the warrant for striking is when the fish has taken out from

30 to 60 yards of line ; gorging will by that period be in nine cases out of ten effected.   Then you tighten the line gently but firmly.   If your reel is supplied with the leather guard before referred to, you press your thumb upon it and upon the spool of the reel ; or if you use the thumb-stall this is the time to test its value.   When the line is taut you may strike, but not very hard.   Some good anglers do not strike at all ; they treat the tarpon as old gorge-bait pike-fishers did the jack.   Never forget that the tarpon if suspicious is marvellously quick in ejecting a bait which has not been absorbed beyond recall.

One of the most glorious sights I know of is the dashing action of a tarpon when hooked.   Frequently, perhaps, in the majority of cases, he shoots perpendicularly out of the water, with rolling eye, and head to sky shaking with fine fury ; when it is a big fish, and this soaring column of silver is 6 to 7 feet high, the effect is verily exciting.   One of the tarpon which I caught leaped at least its own length in the air, not only once, but several times in succession, the water-drops scattering in showers as he rose.

My ambition (let me remark in passing) was to take a series of photographs of the hooked tarpon in its fights for liberty. Wind and weather were against me at the otherwise favourable opportunities, and my success was not what I had hoped ; still, such as it is, it is a genuine photograph of what happens. I tried hard to get my fish as he shot out of water into

"HE WAS ENTERING THE WATER WITH A RESOUNDING SPLASH AFTER HIS AERIAL FLIGHT."

mid-air, but missing that situation, just succeeded in fixing him as he was entering the water with a resounding splash after his aerial flight, having also blown the bait far up the line.

Finding after its leap and fierce shaking of head and shoulders that the hook is not to be got rid of in that manner, the tarpon with mighty power goes off in one rush, perhaps a hundred yards, without stopping.   The wise angler, of

"THE TARPON WITH MIGHTY POWER GOES OFF IN ONE RUSH."

course, lets him go, with a taut line, but without any real attempt to check the express pace.   The guide in the boat meanwhile will be getting in the other lines so as to have presently a fair field and no favour for the hooked fish ; the anchor and pole of the boat must also be. let go.   Although, therefore, you do not attempt to stop the game, you by and by put on all the drag you can with the short rod. You fix the butt in the socket of the waist-belt, the leather protection (either guard or thumb-stall) is pressed with judgment, but firmly against the spool, with one hand, while the line is pressed with the fingers of the other above the reel.

You are now in for probably a smart tussle.   The rod is held perpendicularly by sheer strength ; it is "giving him the butt" with a vengeance, and the top of the little implement, if of the proper make, bends well to the work.   Far away from the boat, as if it were some one else's fish, out comes the tarpon again, glittering, into the air, with its portmanteau-like mouth wide open ; then on falling back with a splash upon the water it lashes the surface, trying ever to get rid of the hook. Sometimes the fish is pretty quickly killed : I have known it done in twenty minutes or half an hour ; at other times, as I have previously stated, the sport may last over an hour, with the severest strain on arms and back during the entire contest. Many an American fisherman employs great force just to see in how little time he can get his tarpon in.

The fish should be brought up to the boat as quickly as

D

possible, for that is a golden rule in all angling.   It will not
be done without frequent rushes, but the experienced boatman will meanwhile have been backing towards and following

"EVERY ONE SATISFIED."

the fish about, paddling gently ; so that you may find yourself
the best part of a mile away from the place where you hooked
your fish before you come to close quarters.   The tarpon when
near the boat, and partially exhausted, has a vicious habit of

burrowing head downwards, and it is not an uncommon thing
to have your boat a second time towed about a lagoon as if
the fish would never be tired out. Even when apparently done,
and the grand game Silver King, by sheer hauling, is brought,
apparently worn out, in on his side, he will renew his flounder
around, or attempt to go under the boat. This is a most
exciting moment, but the short rod enables the angler to frus-
trate the manœuvre without great difficulty.

You now want a sound gaff and a smart gaffer. Many
American fishermen, I noticed, used a strongly-barbed gaff—
an example which I did not follow, though there
may be crises which would suggest it. The gaff, **The Gaff.**
anyway, is thrust underneath the gills into the throat. There
are stories of strong fish breaking the gaff, dashing free from
it, and knocking the boatman backwards, clean out of the
boat, in its violence. Sharp and sure, but not reckless gaffing
is an art, and there are some of the Florida guides who never
make a mistake. The operation properly effected, the boatman passes the end of a rope through the gills from the
outside, secures with a loop, and makes fast to the boat ; then,
every one satisfied, and the angler sitting him down to recover
his breath, the fishing-boat is paddled back to the yacht, the

tarpon towing astern. Arrived at the yacht the fish is triced up.

No one eats tarpon, and the series of photographs, here **The Dead** reproduced, of the catching of a tarpon will indi-**Tarpon.** cate the final disposal of the game to obtain which so much trouble is taken. The yacht has arrived in port, the

"TAKING OFF THE SCALES."

tarpon is landed, and the coloured man in the foreground is in the act of taking off the scales. The fish is weighed and measured, and entry made of the size and circumstances. The operation of removing the·scales is done by the guide. This does not concern the angler much, and as a matter of passing incident, the angler in this particular case (myself) may be seen disappearing up the wharf towards the hotel. The tarpon is not prized for anything but its sport, and after the scales have been stripped off, the carcase is left on the wharf; next

A SMALL ORANGE-GROWER'S HOUSE ABOVE MYERS.

morning it may be covered with vultures, and it is finally taken away to be used as a fertiliser for the orange trees.

Mr. Charles Stewart Davison, a New York gentleman, contributed to *The Spirit of the Times*, at the close of the season in which I took my share, the results of long experience and observation as to tarpon.   He observed that the fish feed up with the flood tide, along the banks of the channel, and

Tarpon Haunts. settle back with the ebb.   As the season progresses, however, they are found farther up, in the narrower parts of the river ; but in March, when the season may be said to begin (though May is the best month for sport), the gorge-bait fishing grounds would be where the river is perhaps a mile or so wide.   I have traced the tarpon forty miles up the river, which would be twenty odd miles above Myers. The favourite haunts are along the edges of banks of the channels, in perhaps five or six feet of water, and here they love to be on the alert chasing mullet.

IN FORT MYERS WATERS

THE reader may not have forgotten that we had arrived at Punta Gorda (see page 9) and found refuge on board a steamer, the *St. Lucie ;* nor that Mrs. Rowland Ward was appointed note-taker in ordinary to the expedition. The following entries from her Diary will be, therefore, a resumption of the narrative which was broken off in order to discourse upon the fish of which we were sure to hear everlasting talk wherever we went ; and I make no apology for using them, because they will indicate very fairly what lady visitors to Florida would be likely to experience. And here beginneth the Diary.

Tuesday, 6th April.—Awake at 5.30 A.M., and up and about by seven, when the steamer left Punta Gorda ; not many passengers on board, but among them an elderly gentleman, a Mr. Plant, the owner of the well-known Plant railroads and steamers in this part of Florida.   He was going to visit Mr. and Mrs. McGregor, who very kindly saw that he caught a big tarpon before he left, which for a man of his years is a great accomplishment.   We had a lovely journey, as there was a delicious breeze.   One cannot see much of the country on such a trip, as the rivers are so wide and the steamer keeps in the middle, except at the ports, where she touches to land passengers, mails, etc.   Passed a lot of small porpoises, which kept right along by the boat; coming up the Caloosahatchie River passed the boats out fishing for tarpon.   Fort Myers looks very pretty from the river—all the wooden houses with their verandahs in amongst the green orange groves, and large palm trees.   Landed about 3 P.M., and after looking at the Hendry House decided to stay at Myers Inn.   Have two nice rooms, unpacked our things, had dinner, and went to bed, very tired and hot.

Wednesday, 7th April.—Got up about seven, and after breakfast put our things straight and looked about the village, which is a funny little place, with a street running even with the river, with a good many stores, which the men seem to laze about outside; in these parts they work very little.   In the afternoon, sailed over and rowed up to Yellow Fever Creek, which is very pretty along the banks, with their mangoes, big palms, palm fans, and large ferns.   Jim[1] fished with the Phantom minnow, hooked five fish, but lost all, which was unlucky.   Saw a Moccasin snake curled upon some rushes; Jim shot him.   We also saw some ducks and water turkey. Lovely day, and in the evening no breeze.   The tarpon boats had to row back.   One gentleman brought home three fish weighing 160 lbs., 123 lbs., and 71 lbs.; they looked very handsome, lying so silvery in the moonlight.

---

[1] "Jim" in these entries means Rowland Ward.

Thursday, 8th April.—Jim was off by 6 A.M. fishing. I had a lazy day, working, reading, and writing on the verandah; it being very hot we could not walk about, but there was a nice breeze on the

A VIEW IN MYERS.

river. Jim returned about 4.30 P.M., having had no luck with tarpon; no one caught any to-day from Myers.

Friday, 9th April. — Jim went fishing at 6 A.M. Lovely, fine, sunny morning, but about 10 o'clock a bad thunderstorm came up, with torrents of rain and a strong wind; the river looked quite rough. Jim returned about 4 o'clock, tired, without having caught anything except two lady-fish. Rained all the evening. Went to bed early, as Jim had bad cold.

Saturday, 10*th April*.—Jim did not start early, and as by 11 o'clock the weather looked bad again, windy and cloudy, he made up his mind not to go fishing to-day. At 2 o'clock it blew hard and commenced to rain, and continued until evening. After breakfast,

YOUNG VIRGINIAN DEER (PHOTOGRAPHED AT MR. McGREGOR'S, FORT MYERS).

walked round village stores and met Mr. McGregor; then walked out to his house and saw the tame deer. Such a nice house and grounds, and they have a yacht in which they cruise and live.

Sunday, 11*th April*.—A fine sunny day, but windy indeed. Jim started about 6 A.M. The guide hooked a tarpon opposite the hotel, but his line broke and it got away. The fishermen did not hook any others. Returned about 5 o'clock. I read and sat on the

LANDING STAGE AT MYERS.

verandah. Mr. Parkinson called in the morning, and while talking to him a black snake walked along the middle of the verandah, but directly we got up and went to look at it, it dropped off and glided away through the grass; they are quite harmless snakes, and as they are believed to kill rattlesnakes, rats, and mice, the people do not destroy them.   But I don't like the things.   Very tired; went to bed early.

Monday, 12*th April*.—Jim off by 6 A.M.   Sunny morning, but

MRS. ROWLAND WARD.

windy.   I had breakfast at seven, then walked up the village to the stores; on returning met Mr. Parkinson, who called to give me a rattlesnake poison fang, which I shall have mounted as a pin, like that which he wears.   Went with him to a "Canning Co." in the village, where they have 25 acres of guava and orange trees, and make guava jelly and preserve the fruit in tins.   There was no ripe fruit, but Mr. Gardner, the owner's son, said he thought they would have a thousand barrels of guavas this year. I tasted one the other day, and did not care for it at all ; it was something like a very poor green fig. On our return we went into Hartigan's and saw some small alligators and skins of rattlesnakes, very handsome ones. Jim returned about 4 P.M. with a fine tarpon, 125 lbs. weight and 6 ft. 3 in. long, a beautiful silvery fish ; sent it to Hartigan's to be skinned. Went up the village and on *The Tarpon*, which arrived about four, and ordered in our stores, etc., to make an early start in the morning. Terribly windy day.

OUR LIFE ASHORE AND AFLOAT

WHILE entry by Diary has its advantages when it is written from a lady's point of view, and should therefore not be tampered with (as the foregoing has not been), there are drawbacks. For example, the Diary states that I lost five fish, "which was unlucky." In truth it was due to the hooks hanging loose (I afterwards rectified this) from the Phantom minnow, and the snags. That day I saw numbers of ospreys, darters, buzzards, coots, and other birds. As to Punta Gorda, when I came on deck in the morning previous to our start I found a white mist veiling all the prospect, mullet splashing all around, and flocks of duck (scaup) sitting in happy confidence on the smooth water.

At Fort Myers Inn my first impression was that it was all verandahs ; and one evening (it must have been the 9th)

THE BIRDS HERE ILLUSTRATED ARE THOSE USUALLY SEEN IN FLORIDA WATERS.

the fireflies were splendid, floating through the rainfall.   Those

verandahs were certainly a fine institution when you wanted

to get away to write, read, or meditate.   I may

state, too, that on the 8th April Mr. Holloway, **Life Ashore.**

a noted tarpon fisher, caught three tarpon, one weighing

163 lbs.   On that day I saw another angler losing a tarpon

after it had towed his boat about a mile and a half, and

jumped out of water several times.   The catastrophe occurred

quite close to my boat, and the angler had lost three pre-

viously on the same day.   In the morning I was startled

while reading by the noise of wings, and looking up saw

some two hundred ducks within shot.

At Fort Myers, just where the Diary on a preceding page

is interrupted, a break occurred in our method of living.   We

left the inn and took up our quarters on board an auxiliary

naphtha yacht, named *The Tarpon*, on the 11th April.   I

chartered her for 22 dollars a day, and put in her about five

pounds' worth of naphtha, which is very cheap there.   Besides

this the crew have to be provided with food.   The reader

may see what she was in appearance by the illustration.

As to the living ashore, perhaps the fairest opinion I

can offer is that it might be better and it might be worse.

E

In an out-of-the-way corner of the world like this you cannot expect the luxuries of a great city, or a fashionable watering-place like Tampa, higher up the coast. The style of living at the inn was, generally speaking, the American

YACHT "TARPON."

fashion, which surrounds you with a number of little dishes; but it is no use pretending that the meat was otherwise than tough, and as it is produced from the lean kine of the adjacent woods, and has to be cooked very soon after killing on account of the heat, its quality may be readily imagined. There is, however, an abundance of good fish, but the most acceptable dishes were the turkeys and the chickens, though they were not on the *menu* as often as they might have been.

The chief meal (dinner) was at 6.30, and the vegetables consisted of ordinary and sweet potatoes, white French beans, and tomatoes. Corn bread was the general staple, and they gave us sweets, with either iced tea or hot coffee. The cook was a black man, so also was the waiter, who, besides looking after us at meals, did all sorts of things, I believe, such as milking the cow, feeding the chickens, and meeting and seeing off people by the steamers. It is true there were no baths at the inn, but good-sized washing tubs were provided as a sort of compromise. I must say that the host and hostess did the very best they could for their visitors, and were unceasing in their attention. Probably by this time there is a large hotel built at Fort Myers, with all the improvements and attractions for visitors which I understand characterise the hotel at Punta Gorda, which we found closed. Our charges at the hotel were two and a half dollars per day, or fifteen dollars a person per week. Wine, beer, or spirits you must provide for yourself;

and if you want good whisky take it from the Windsor Hotel, Jacksonville.

Cruising. The yacht, on which we made our home on leaving the hotel, was a little vessel between 20 and 30 tons. There was a main cabin used for a dining-saloon and drawing-room, and its centre-board top made a tray, on which stood the iced drinking water, with its usual tap, and various odds and ends. Two beds were made up on the cabin cushions, and we were supposed to be protected from mosquitoes and other insects by a gauze screen, which, as usual in such cases, proved ill-fitting, and admitted the few pests which were amply sufficient to interfere with our comfort.

We had much better food on board, for our little Jap cooked admirably, and we took care to lay in plenty of stores, having arranged for a supply of beef and mutton from Jacksonville. This meat came down by steamer in a box, packed in ice, and all we had to do was to send our boy in the yacht's boat to fetch it. In this way we also got chickens and venison. The prices paid for these welcome contributions to our larder were—venison, $12\frac{1}{2}$ cents per lb.; eggs, 25 cents per dozen; chickens, 35 cents each; turkeys, 1 dollar; milk, 10 cents per quart.

Our best meat was the venison, which was very tender.   I
believe it was the close time for deer, but somehow we had
no difficulty in purchasing a supply.   We had delicious tomatoes
and French beans, and we could get any quantity of these
wholesome vegetables at low prices.   Indeed we had them for
nearly every meal, and thus by the aid of omelettes and a sort
of pancake, in the composition of which our clever Jap *chef*
was a real artist, we did fairly well.   How he managed his
work was a standing wonder to us, as the galley at which he
had to do everything was only a few feet square, and his fuel
apparatus was the three fires of gasoline making big flames,
with usual fittings for cooking utensils.

There were, of course, on board drawbacks of no inconsider-
able character, but as we survived them all, I will not enter
into details as to the monster cockroaches which used to run
over us, and the leaks which soon forced their attention upon
us.   These boats being uncoppered, the hull has no protection
from the destructive worm which bores into the wood ; and it is
not pleasant when you get up in the morning to put your feet
into dirty water which has come in by leakage during the night,
and to have to get the pumps used daily.   Naphtha launches
are used a good deal in these waters, and I saw no accident

nor heard of one with them.   At the same time, they require
very careful handling; and as an American said to me years
ago in New York, "They are all right *in some people's hands.*"

DODE, THE COLOURED GUIDE, SITTING ON COCKPIT, HOLDING TARPON LINE.

As I went to America long since to see this type of engine,
I perfectly understood what was meant.   The fact is, naphtha
is a more dangerous spirit than most people ever really imagine.

I should recommend, however, a sloop as the best form of
sailing-boat for cruising on the fishing grounds—such a boat as

is depicted on another page (p. 10). It is much cheaper, and can be hired at reasonable prices. You can engage a man with sailing boat and small boat at, say, from 5 dollars a day.

OFF THE MAIN RIVER, ABOUT TWENTY MILES ABOVE MYERS.

It is not necessary to follow my example and pay 22 dollars a day.

A man wishing to be very economical, and who has no objection to living under canvas, may set up his tent on a place like Captiva Pass, about which something will hereafter

be said.   He would then only require a rowing boat and a
man for fishing purposes, and if only intending to fish for
tarpon, two rods and lines and one reel will be sufficient.
There is generally a launch or sailing boat at hand to tow
him to the grounds, or he may have his lug sail for his
own open boat.   He might cook for himself, or utilise his
boatman.   As there is a steamer passing daily, there is no
reason why, wind and weather permitting, he should not
get away as far as five miles for choice of fishing grounds.
The insect pests, however, would be a consideration as the
season advanced.   Stores he might procure from Jacksonville.
At the time of going to press I hear that Captiva Pass now
has a floating hotel—not a bad idea at all.   It was constructed
and is conducted by Mr. J. B. Hughes of St. James' City.

THE SHEEPSHEAD.

VI

PASS FISHING

THE expression " Pass Fishing " which you hear so frequently amongst the regular fishing visitors to the west coast of Florida, the guides, and the residents, is both geographical and, in a measure, technical. It at once indicates a special fishing ground and a style of angling which is different altogether from the commoner. practice, in pursuit of which the angler sits meekly in his boat, praying humbly and waiting meekly for a bite.

The place is the Captiva Pass, spoken of sometimes as Captiva, but more often as " The Pass." As a **Captiva** reference to the map shows, it is off the west coast, **Pass.** a channel between two islands ; and the tide, in and out of the

Gulf, according to the laws of ebb and flow, spring and neap, runs at from 5 to 8 knots an hour. The breakers outside are no doubt picturesque and musical in their roar and murmur, but they are mighty in power, and always gave me the impression that, although this may be and is an excellent place for miscellaneous sport, it is not exactly the spot in which one would care to be capsized. In fishing, however, the boat is anchored well clear of this apparent danger. Sometimes at a push it is necessary to have your boat on the move as quickly as possible, and it is the custom therefore to attach a buoy to your anchor, and so, at the critical moment, it becomes simply a case of casting adrift from your moorings and picking them up afterwards.

The style of fishing here is commonly called trolling, but **Trolling or** it is not at all that according to our English **Trailing.** acceptation of the term. Perhaps it may be best described as automatic trailing. It differs in most essential respects from gorge fishing for tarpon. In the Pass you sit with your rod in hand, with some 30 or 40 yards of taut line out, and everything free for the running of the reel when a fish strikes. The current works the bait briskly enough for all practical purposes. A large spoon or an artificial minnow would of course be serviceable at this work, but as you are always likely to get hold of tarpon, bass, and other big fish, it would be necessary to reckon upon an enormous loss of baits. It is in consequence the common practice to bait your single hook or triangle with a thin strip from the side of a mullet.

There is very seldom any mistake, in one sense, about Pass fishing, for the tarpon and other species when they are on the feed dash vigorously at the moving bait. They prowl around within a few feet of the boat, and take the bait almost before it touches the water. The hours sometimes are very lively, but the percentage of misses is enormous. In the case of tarpon the bony mouth explains the cause, and if you get one fish in ten you may think yourself very fortunate. It should also be remarked that a very strong reason for fastening your boat to moorings rather than an anchor, as at Fort Myers and up the river, is the likelihood there is of a big fish, on being hooked, making directly for sea as hard as he can. Yet it occasionally happens that even in this style of fishing in the Pass there is no sport, no sign of fish, and then it is the proper thing to cast yourself free and pull slowly about, adopting what is, to all intents and purposes, the harling method which we all know so well in Norway and Scotland. If you hook a monster when so engaged, the sport is prolonged at will ; your man can take you leisurely ashore, and gaff the fish at his ease in shallower water.

It always occurred to me that an angler who was content to see his baits carried frequently away, and equipped himself now and then with ordinary spinning tackle, might have a real glut of sport in this style of fishing, casting from right and left, and experiencing sensations and excitements, along with smashes and losses, such as could not be exceeded anywhere.

On another page I have made allusion to Mrs. Grimshaw. She is the lady who, in the *Badminton Magazine,* described

Mrs. Grim- her experiences of tarpon fishing, and of Florida
shaw and generally.   She pictures Captiva Island as a very
her Sport. beautiful spot, its beach sparkling with myriads of shells, and its semi-tropical growths of prickly pear and palmetto, all charming.   Indeed, she describes the spot as enchanted, and was evidently much in love with it.   She stopped in one of the huts inhabited by the Spanish fishermen, but in 1897, I fear, she would have had a very hard time, for the place was infested with mosquitoes and sandflies, which at times made life almost unendurable.   The fisher people, who supply you with bait and are very obliging to the visitor, told me that there had been

none the previous year, so Mrs. Grimshaw was very fortunate. The lady's first impression as to the general scenery of Florida was similar to our own.   She was struck by the monotonous dead levels, and the succession of swamps and lagoons, but

TWO DAYS' CATCH BY A LADY AND HER HUSBAND AT CAPTIVA PASS.
Jewfish on the left estimated to weigh 795 lbs.

she saw a certain charm in the groups of palmetto trees, the brilliant masses of creeping flowers, and even the murky-looking bogs and lagoons, with the possibilities of alligators and rattlesnakes.

In the matter of sport, Mrs. Grimshaw's experience in this

Pass (which is ¾ mile wide and ½ mile long) is that the likeliest haunts of the fish are the tide rips.   When there are a number of fishermen out, and each is striving with amusing eagerness to get the best anchorage ("swim" we should call it at home), the sport is intensified by the necessity there is of the boats looking after one another ; for it is an unwritten law amongst Florida sportsmen to clear out at the moment of action, and not get in the way of another man's fish ; and when, as I have seen, three or four men happen to be engaged simultaneously with a fish, the fun is fast and furious.   The raw hide snell is, in Pass fishing, exchanged by many for a length of piano wire, which will hold anything ; and the same rod is used as in gorge-bait fishing.

The number of misses in proportion to strikes is illustrated by Mrs. Grimshaw's statement that her husband once counted **Fine Sport.** seventeen strikes in thirty minutes to his own rod, without one fish being hooked.   It did actually happen to her that, hooking a big fish which came up four times, always in a fresh place, shaking his head frantically, her guide promptly flung out his buoy, and the boat was at once towed with a seven-knot tide towards the Gulf.   This was by a hooked tarpon, and the course I have

described of pulling gradually in towards the beach was fortu-
nately adopted. The boatman jumped into the knee-deep water,
Mrs. Grimshaw scrambled after him, and waded ashore, and
it was then that the real tug-of-war seemed to commence.
The lady angler confessed that it was no joke running up the
shingly beach, as she did, with 130 lbs. of fish fighting for dear
life at the other end of the line. The battle is most vividly
described in the *Badminton Magazine;* victory came, the fish
was duly gaffed, weighed at 125 lbs., and measured at
6 ft. 3 in.

The next night her husband hooked a gigantic fish which
towed him quite into the Gulf, and he and his guide
disappeared for two hours, were searched for, and found
triumphant but exhausted after a fight with a tarpon weighing
175 lbs. Mr. and Mrs. Grimshaw were in great luck at that
time, for the same night the former caught a second tarpon
weighing 150 lbs., and she herself another of 135 lbs.; well
might she write, "they did look three beauties lying side by
side in the moonlight."

Reference will be noticed in the Diary to Mr. Mygatt.
There is no better known name in these waters than his, and
it may be remembered that he also contributed two articles

on tarpon fishing to the *Badminton Magazine* in 1895.   He has fished the same waters on successive seasons, and is

Mr. Mygatt's Experience. therefore fully qualified to speak.   His first season left him a decided pessimist, for his luck was wretched ; but after his second and subsequent seasons he became, and remained, what he calls a "rabid optimist on the sport." That was a grand day of his when he caught eight tarpon of 155 lbs., 138 lbs., 128 lbs., 139 lbs., 136 lbs., 110 lbs., 119 lbs., and 78 lbs.   With the exception of the last, which was 5. ft. 10 in. long, all these fish were considerably over 6 feet in length, one of them, and that not the heaviest, reaching 6 ft. 8 in.

During the same season he had other enviable days, and his heaviest fish was 182 lbs., and at that time it was, as far as Mr. Mygatt knew, the longest of which there was any accurate record, for it measured 7 ft. 4 in.

According to this gentleman's observation the tarpon frequent the shallow lagoons and the brackish and often sheltered **Tarpon Seasons.** water of the adjoining creeks from September to June, but during the rainy season, viz. from June to August, they seem to abandon the creeks altogether and also the upper lagoons ; he imagines that this may

be accounted for by the combined reasons (1) that the
water becomes too sweet for the fish ; (2) that they go out
into the Gulf to spawn. From my own inquiries and the
answers I received to my questions I should imagine that there
is no very authentic information as to the times and manner
of spawning, but Mr. Mygatt states distinctly that he never
caught any fish in roe before May. He confirms my
opinion that the best months for catching tarpon are April
and May in the spring, but he also adds that October and
November are good months. For the angler who wants to
have real sport, concurrently with pleasant weather, April
and early May are preferred. I notice that Mr. Mygatt,
like myself, would also like to hear of a specially effective
hook, devised to penetrate and grip the hard interior of the
tarpon's mouth. He believes that then the fish might be
taken with a large fly.

It has been said on a previous page that small tarpon may
be caught with a fly, and that Mr. Parkinson, an English
gentleman, actually did obtain the record for the **Artificial**
smallest specimen of tarpon with one caught with **Tarpon Fly.**
that lure. Mr. Mygatt narrates how he took his very first
tarpon with a huge fly made by himself. He was up the south

F

fork of the St. Lucie River, using a channel bass rod, trolling and casting for crevalli, with a spoon bait, when a tarpon seized the hook, jumped, and escaped scot-free. All that day he continued striking fish and losing spoons and phantom minnows. In one tarpon school he thus had nearly fifty strikes. He returned sadder and wiser to his sailing boat, and spent hours in dressing "most complex and weird jumbo flies"; returned to the spot next day, and found what he thinks was the same school of tarpon, which had advanced by this time some two miles farther up stream.

His flies were made of pieces of wood 3 inches long by $\frac{1}{2}$ inch in diameter, covered with red flannel, and tied with three or four feathers. Each was about 4 inches long, and terminated with three bass hooks, set at different angles—a lure, in fact, very much like the flies we use in England for pike fishing in shallow lakes. There was no end to the rises at this delicacy. The angler repeatedly struck hard enough to make the fish jump out of water, but they always managed to get rid of the fly or sever the snell. One fish at last did take the fly and the hooks, and after forty minutes' battle was gaffed. This was Mr. Mygatt's first tarpon, but it had been taken by one of the hooks entering the skin that covered the head. There must have been a bit of smart angling here, as the hook had ripped along the skin for about two inches, and the hold was so precarious that the fly fell off as the fish flopped about in the boat.

Tarpon fishing at night is recommended by Mr. Mygatt. He declares it if anything more exciting than in the daytime, since to "the many usual details are added the indescribable fascination and mystery of darkness. To catch **Night** tarpon at night by moonlight is pretty hard to **Fishing.** beat for sport, but I must say that to catch them on a pitch-dark night when all the playing has to be done by the feeling of the strain on rod and line is the most exciting sport I know of. I certainly prefer night fishing. It is always cool, there is rarely any wind, and if not late in the season, not many mosquitoes."[1]

At Captiva Pass you fish best on the flood-tide, and it was while here that Mrs. Ward and myself lived on board the yacht. We arrived there on 14th April, after some **At the** "small fishing," and reached our anchorage in time **Pass** to have an hour or so of fishing before dark. Here **Anchorage.** we soon had strong evidence of the obdurate character of the

[1] "Some Tarpon Adventures," by Otis Mygatt, *Badminton Magazine*, October 1895.

interior of a tarpon's jaws, for we lost a number of fish, and the other boats that were there seemed to be having a similar experience. Mrs. Ward lost two fish in succession, after she had played them some time and certainly deserved to bring them to land.

Sharks are a great nuisance here, and it is no uncommon thing to hook your tarpon, and after holding it a while, the line comes back, and you find a couple of feet taken

Sharks.

off the fish at one fell swoop by one of these predatory sea-wolves. In still, shallow water, sitting one day in my boat angling for small fish at the back of an island and over an oyster bed, I saw a shark sheer alongside the boat, punting itself lazily and unconcernedly along, and it certainly was not less than 14 feet long. On the point of Captiva Pass at the time of our visit was the camp of Mr. Von Blake. He caught a large number of tarpon, and the season before our visit, I believe, he had achieved the very fine record of 70 fish. We were not the only English visitors : there were several English gentlemen there, doing well ; and one of them told me he had had ten strikes, but had lost every fish.

The Pass, at certain times of the tide, seems to abound with fish of all kinds, and with their antics on the surface resembles nothing so much as a boiling cauldron. While you are fishing, the tarpon are on the move all around, and sometimes they are so numerous that they might be taken for a shoal of gigantic mackerel. There is no shyness in these uneducated fish : they come **Miscel-laneous Sport.** near the boat ; indeed they sometimes jump right over and even into it. The tarpon, which in certain lights show a bright blue edge to the scales, look most charming in the water with their light green backs and silvery sides. Turtle, although very shy, come up and go down in the Pass alongside your boat, and having shown their Venetian brown heads, silently steal away. Some of these I should imagine would weigh at least 400 lbs.

The beautiful Spanish mackerel (*Scomberomorus maculatus*) is caught here ; it has shades of gold all along the side, curiously marked, and suggesting that they have **Spanish Mackerel.** been smudged on with a human finger. One day I saw a gentleman catch 35 mackerel with his fly-rod and artificial fly. My fly-rod was of course left behind, but I furbished up and put a white fly on the tarpon tackle, and caught a number of mackerel with this primitive form of fly-fishing.

There is another species, known as the lady fish. It is the same species, I think, as that called the bone fish, or grubber, in

The Lady Fish. the Bermudas. It is the *Albula vulpes,* and it is so primo a game fish that it elicited the praise of Dr. Henshall, the famous American angler-author. He had been catching in quick succession " salt-water trout," as the squeteague are called, a few red fish or channel bass, some ravallia and crevalli, but his 3 lb. lady fish gave him more real sport than any of the others. The specimens I caught here were certainly remarkably sporting. Slender and silvery, they fought as well as the black bass, and leaped from the water—not going straight upwards, however, like a tarpon, but shooting out right and left, as if they wanted to learn the tricks of the flying fish.

Mrs. Ward celebrated her Good Friday inside the islands of Captiva Pass by catching a large number of channel bass, and so-called trout, in the slack, shallow water. My remembrance of the place is principally of fishing I had by moonlight, when the heat and mosquitoes gave me the agony of swollen feet and general discomfort that were very trying. I lost a very decent tarpon on this ground, but went out next morning at six o'clock and got 20 lbs. of weak fish, two bass weighing 9 lbs. each, and eight lady fish averaging 14 lbs.

CHANNEL BASS.

VII

CONTINUATION OF DIARY

WE may now return to Mrs. Ward's Diary, taking up the record at Fort Myers, from which I have somewhat strayed, by anticipation, to explain the different style of fishing in the Pass.

Tuesday, 13th April.—Hairdresser came at 6.30 and shampooed my hair; had breakfast, and about nine we were ready to start on *The Tarpon*. When we left the hotel dock, large buzzards were all round it after a tarpon which had been thrown down there and was not wanted. There was a fine strong breeze and we sailed well. Passed all the little boats, tarpon fishing, and about 11.30 arrived at Red Fish Point, where we dropped anchor and put up the awning. We had lunch and then started fishing; Jim and his guide in one boat, and the Captain and I in another. I caught a Spanish mackerel and missed one or two strikes. Jim caught two "trout" and twenty other fish. This was not considered good, as the tide was wrong and there was too much wind. Still, very peaceful and nice on board.

Wednesday, 14th April.—Up at 5.30 A.M. After coffee, started off fishing in the boats soon after six. A beautiful calm morning, and the sun shining; but no luck followed me, and the Captain and I only had one strike and caught nothing, though we trolled for two hours before returning for breakfast. Jim caught some "trout." After breakfast, at about 10 A.M., we started for Captiva Pass; there being little wind, we steamed as well as sailed. Passing Punta Rassa, the

PUNTA RASSA CABLE STATION.

men hailed us and gave us a cablegram for Jim. Already they had learned to identify us by the boat.

Arrived at Captiva and dropped our anchor soon after 4 P.M. A few minutes later we were both out in the Pass fishing. I caught nothing, but Jim hooked a tarpon, which jumped finely and got off.

A very hot day ; could not stay on deck.   Coming along from Red
Fish Point, the heat was great, but now it is deliciously cool under the
awning. We are only about 100 feet from the shore of the island,
which has a beach of nothing but small shells ; a few Spaniards live
on it in huts made of palmetto grass.   They are fishermen and catch
the mullet in nets.   The water is alive with cat-fish ; when anything is
thrown overboard, they generally come round in hundreds and fight
for it.

Thursday, 15*th April.*—Up at a quarter to six and had a cup of
coffee, then started out fishing.   I went over to the other side of the
Pass and along the shore of the island ; trolled and caught thirteen
squeteague weighing 17½ lbs., one bass 6½ lbs., and two bass 5¼ lbs.
Returned about 9 A.M. to breakfast, and found Jim had only caught
six squeteague.   Nice fine morning with strong wind.   The
mosquitoes are very tiresome, and last night bothered us a good
bit ; the Captain had to come up and sleep on deck, they were so
bad in his cabin.   Jim caught two red grouper and one black grouper
in the morning.   Started out to the Pass again about 3.30 P.M.
Heaps of tarpon playing about in the water ; I hooked two, but after
two or three jumps they were gone ; they were very big fish and
the pull was tremendous.   I fear I should never have strength to
play and land one, but must try.   I saw only one gentleman catch a
tarpon this evening, but several had strikes.   Went in to dinner at
seven, and early to bed.   Terribly hot night ; Jim could not sleep
at all with the heat and mosquitoes.

Friday, 16*th April.*—Lovely sunny morning.   Up at 5.30 A.M. and
out fishing early.   Caught three squeteague and a small "jack" ; seeing
a shoal of mackerel, went and trolled among them and tried all kinds of
bait, but could not hook one, though the water was alive with them.
Landed at the island at Mr. Von Blake's camp, and saw two lovely

tarpon he had caught this morning at 4 A.M. One weighed 165 lbs. Several other tarpon were hooked, but a shark mangled two of them, leaving only the head of one.

Jim had good luck before breakfast, for which he returned rather late, having caught two bass of 9 lbs. each, fourteen squeteague weighing 20 lbs., and eight lady fish weighing 14½ lbs. He had thrown several lady fish in again and lost a very big bass. About 4 P.M. he went out into the Pass after tarpon, and hooked one but lost him. Cooler towards evening; the wind got up, and it became very cloudy. The mosquitoes seem less plentiful; last night they were terrible.

Saturday, 17th April.—Under weigh about 6.30 A.M. Wind having all dropped, had to steam. Stopped at St. James; went ashore and bought some bread and eggs and other things at the store; then walked up to look at the Don Carlos Hotel, a nice building, with only one gentleman staying there. In the garden, tied to a tree, was a small alligator, about 6 ft. long; its mouth was bound up with string, it made a most funny noise, and would have lashed one with its tail if you had gone near enough. Steamed up to Red Fish Point, dropped anchor about three, started to fish for tarpon, but had no luck. Wind got up in the evening, and it blew very hard in the night. It had rained off and on all the morning, and was very cloudy.

Easter Sunday, 18th April.—Still at Red Fish Point. We started fishing off the boat about seven, before breakfast. Had no luck, only hooked two small sharks. Fished all day off the boat, but caught nothing. Jim rowed down the river a little way and caught some squeteague and lady fish. Fine day with strong wind.

Monday, 19th April.—Saw no tarpon about when we looked out; therefore determined to go up the river, and at 6.30 started and steamed three miles above Myers. Jim went out in the row-boat, but seeing no fish about, we started down the river and I went off the

yacht to the inn. Jim, however, went out fishing again, but soon had
to return, driven home by a terrible wind-storm with torrents of rain,
and thunder.

Tuesday, 20th April.—Jim up at 5 A.M. and out fishing by six;
caught a tarpon 114 lbs. weight opposite Mr. McGregor's, the other side

MR. McGREGOR'S SCHOONER BELOW MYERS, OPPOSITE HIS WINTER HOUSE.

of the river, about 3 o'clock ; landed at the McGregor and walked home.
Fine, but very windy day.

Wednesday, 21st April.—Very windy. Jim did not go fishing and
had a quiet day.

Thursday, 22nd April.—Another windy day, and as it blew hard
all day, none of the gentlemen went out fishing. Mr. Batley arrived

about 7.30 in the morning with a pair-horse hack, and we drove nine miles out in the country to see some fine orange groves. We travelled through ground thick with palmetto green and fir trees, except where it had been burnt down or cleared for an orange grove. The road was in places very sandy, and we went through streamlets and across wooden bridges over the creeks. Some of the orange groves were fine big trees, twenty years old, and others quite small ones just planted; between the trees, sugar cane or tomatoes were growing. The tomatoes are ripe now and being shipped daily in boxes to the northern markets. On our way back, we stopped at the experimental station supported by the State, and saw some acres of pineapple, but none of them ripe. There are twenty varieties growing there, and all kinds of plants and flowers. The man in charge gave me a beautiful crimson hybiscus bloom and two or three roses. Got back about 12.30 to the inn.

Friday, 23rd April.—Lovely sunny morning with wind dropped a little. Jim took a black guide, named Dode, and started off fishing about 8 A.M., returning about 5 o'clock with a tarpon weighing 41 lbs. Mr. Van Cortlandt got one also, weighing 62 lbs.; none of the others staying here brought any home. They hooked several but lost them.

Saturday, 24th April.—Fine morning, but still windy. Jim started off about nine, fishing, and returned about five with a tarpon of 92 lbs.; he was the only visitor at this inn who caught one.

Sunday, 25th April.—Lovely morning with wind dropped. We were up at six, and at seven started off up the river in the little steamer, the *Belle of Myers*, which we had hired for the day. We voyaged 18 miles to a place called Telegraph Creek, where we stopped for two or three hours and fished. It was very pretty going along the river, which gets narrower as we advanced. We saw five or six alligators, which dropped quietly down into the water when they heard the boat, and were lost to sight.   One small fellow was swimming across the Creek in front of our boat.   We caught about twenty fish, mostly small bass.   Stopped on our way down and fished a creek; reached Myers about six.   The wind had by this time risen, and was blowing hard from the west in the evening.

Monday, 26th April. — Lovely morning.   Went out with Jim,

THE "BELLE OF MYERS."

fishing, about seven, and returned about 3 o'clock, but never even saw a tarpon.   Mr. Jones caught a tarpon weighing 187 lbs.

Tuesday, 27th April.—Jim up and off by 6 A.M.   Lovely day, very hot indeed and calm; he never even had a bite, though he saw lots of tarpon.   Mr. Holloway caught four beautiful fish and Mr. Van Cortlandt one.

Wednesday, 28th April.—Jim off by six.   A lovely morning, and still very hot.   He returned about five with a nice tarpon of 122 lbs. Mr. H. caught two of 162 lbs. and 157 lbs.; and Mr. Van C. two just over 100 lbs. each.

Thursday, 29th April.—Dull morning, with more wind. Jim did not go out early, but after breakfast went up the town, and started about nine fishing. Blew up to a strong gale, and river got very rough. The fishermen all returned early about three with no tarpon.

Friday, 30th April.—Woke at 2.30 A.M. by a bad thunderstorm, which lasted until eleven with pouring rain. No fishing.

Saturday, 1st May.—Jim went out fishing.

Sunday, 2nd May.—Jim up early and went out fishing. A lovely day ; he returned about 4.30 with a large saw-fish on his boat which Mr. Vom Hofe had caught ; he had it on for over two hours before he could see what it was ; it was 14 ft. long and 4½ wide, and weighed about 600 lbs. We all photographed it.

Monday, 3rd May.—Up at 4.30 and had breakfast ; went on board the *Clara* to start at six ; proceeded to Punta Gorda, thence by train to Lakeland and Jacksonville.

There is really very little in the shape of narrative to add to the entries of this Diary. Lady readers will, no doubt, have

Ladies in Florida. been interested in reading of the sport which Mrs. Ward herself enjoyed, and if a lady can endure the heat, mosquitoes, and a certain amount of roughing it in Florida, she might pass a very pleasant time with a lightish rod and tussles with a variety of fish. I managed one day to get a " shot " at the good keeper of our Diary, but it was not under favourable circumstances. The overpowering sun on that May day proved too much for the details one would have

liked to emphasise.   This was towards the end of our visit,
and the angler, as seen, has been having some fair sport in the
river, very near the spot where Mr. Parkinson caught his small
tarpon with a fly.   She is really engaged in putting on a

FISHING IN THE CALOOSAHATCHIE RIVER, TWENTY MILES ABOVE MYERS.

phantom minnow to replace one which has just been rudely
snapped off by a predatory bass.

Tarpon fishing is, however, very severe work for a lady,
but it must be remembered that the record amongst anglers of

both sexes is the tarpon of 205 lbs. caught by Mrs. Stagg. We saw her tackle on board the yacht, and in the cabin were photographs of five fish caught by her in two days' fishing in the Pass. One of these tarpon had the best part of two feet bitten off by a shark.

As to all that has been said about mosquitoes, although perhaps the references to these little pests have been too many, they made such an impression upon me that I am bound to say the facts have not been at all overstated. They always made very free with my ankles, and on one occasion they had bitten me so furiously that I was absolutely unable to sleep, and had to keep my poisoned and swollen feet and ankles in iced water for a while. The sandflies are also, as in other parts of the world, worse than the mosquitoes, both from their numbers and their minute size.

I see no mention made in the entry of 27th April of Mr. Ashby Jones's heavy fish. This was the same gentleman who

Mr. Ashby Jones's big Tarpon. a fortnight before hooked a large tarpon which towed his boat about a mile and a half, jumped several times, and escaped, not very far from our own boat; and he had also lost three on the previous day. During dinner on the 27th April we heard that Mr. Jones had

killed a tarpon of 201 lbs., and it was stated during discussion
that this was 4 lbs. less than the record fish caught by Mrs.
Stagg.   After dinner, therefore, I went with Mr. Van Cortlandt
and other visitors to see the monster.   It measured 6 ft. 11 in.,
and girthed 41 in.   Without question it was the finest fish I
had seen, and by the light of the lanterns it looked really
bigger than it was.   Talking over the measurements, and not
being quite able to make them tally, the fish was reweighed by
all the scales which could be borrowed in Myers.   These weighing machines, being those of sellers and not buyers, varied a
good deal, but we ultimately agreed to book this fish at 178 lbs.

The following clipping, from an American paper, is the
story of the smallest tarpon already incidentally mentioned :—

Mr. A. T. G. Parkinson now holds a tarpon record.   He hasn't
beaten Mrs. Stagg's record of 205 lbs. for the largest fish, but he has
gone to the other extreme and caught the smallest **A Baby**
tarpon on record.   The baby tarpon was taken with a **Tarpon.**
fly at Alva by him on 21st January last, and weighed
1½ lbs., length 18 in.   Last week he was fishing for bass at the same
place, using his rod and reel with a phantom minnow, when he caught
another baby tarpon weighing 2 lbs. and measuring 19 inches.   The
scales upon which the records are placed are not larger than a cent
piece, and make quite a contrast alongside the large three-inch scales
of the 100 lb. tarpon.

G

Amongst the scraps which I cut out of the papers during our sojourn in Florida was one which I here quote, not as an incident of sport, but as a curiosity of river scenery at Fort Myers :—

A Lake of Fire.

Many northern tourists go to Nassau to see the lake of fire or phosphorescent lake, for which they pay the privilege fee of forty cents. Now the phenomenon is so common at Fort Myers that it is seldom ever thought of, and it is treated so passingly common that the attention of strangers is rarely ever called to it, when in fact it would be one of the greatest attractions to them. The waters of the Caloosa-hatchie, during the darkness of the night, are not infrequently seen glowing with a phosphorescent light ; with every movement its waters burst into brightness, and the refulgent waves appear like billows of fire. The phenomenon is one of great beauty and magnificence. While rowing a boat on a calm, dark night there will be a faint, delicate light dripping from the oars, while, apparently, a streak of fire will follow the boat. Fish can be plainly seen in the water, which makes them easy victims of the "grains," and furnishes great amusement for the lovers of piscatorial sport. Steamers plying the waters often drive before their bows two billows of liquid phosphorus, and in their wake they will be followed by a milky train, which makes a seething and hissing noise. If in the dark it presents a splendid exhibition of aquatic fireworks, looking as though a thousand living rockets were seen shooting from the steamer in all directions, and whirling about in flame-like paths, till the whole river will seem medallioned with fire.

SOME MONSTERS

IN the verandah, notably Murderer's Row, at Punta Rassa, the evenings are whiled away over the cigars with marvellous stories of monsters. There was much talk of a certain Devil's Ray that is 20 feet across and weighs any- **Fish Yarns.** thing up to a ton. These creatures are to be taken by the harpoon, and there were samples of that form of tackle on board Mrs. Stagg's yacht. From the few clear statements I could gather I conclude that this undoubtedly must be *Ceratoptera vampyrus*.

As to tarpon, my biggest of 128 lbs., and Mrs. Stagg's of 205 lbs., were mere dwarfs compared with the fish which was said to have leaped on board a steamer between Fort Myers and Punta Gorda, and smashed in the galley door as if it had been made of cardboard. Another yarn was that an angler hooked, played, and lost a tarpon in the river,

8

near Myers, and that the identical fish was caught within four-and-twenty hours in a different river seventy miles distant. These and similar stories never lose in the telling, and they are always well told and most diverting.

Personally I can vouch for a very large saw-fish. It was the very biggest specimen which I have seen caught with **A Huge** rod and line, and I had the privilege of assist-**Saw-Fish.** ing in its capture. The photograph was happily very successful.

We were out one day gorge-fishing for tarpon, and Mr. Vom Hofe was anchored not far from my own boat. I happened to look up just as he had hooked a fish; that it was no small fry was soon evident. It was running deep down, and there was no leaping in the air or breaking the surface. It towed the boat about for nearly an hour and a half, and I set off to render any assistance that I might. The boatmen had decided, from the working of the game, that it was a saw-fish, and I was able to photograph some of the closing scenes. I put a man on board Mr. Vom Hofe's boat, and all in good time the monster was brought alongside and secured by the adroit slipping of a noose over the ugly saw, which was most conveniently fashioned by nature for such an operation. A

MR. VOM HOFE'S SAW-FISH.

mighty blow between the eyes with an axe then settled him, and the next business was to get him home. The illustrations indicate the manner in which the cargo was moved and secured. Ultimately we laid the fish across the bows of my sailing boat, and I had to beat back to Fort Myers, a matter of miles against a head wind and rough water. The voyage was made all the more difficult by the weight of the saw-fish, which often caused the nose of the boat to dip inconveniently into the waves. At Fort Myers we transferred the fish to a smaller boat (see illustration), and then came the hauling of it up to the verandah, where it was photographed, with the proud captor by its side. It will be noticed that his rod and tackle were those ordinarily used for tarpon, and wonderfully light for such heavy work. The saw was cut off for preservation, and the carcase left upon the wharf. Thither came all and sundry armed with knives wherewith to cut off strips of skin to keep *in memoriam.* The length and weight of this monster are stated in the Diary.

It was not my privilege to see a still larger fish, but I **A Huge** heard of it, and was presented by an American **Jew-Fish.** gentleman with a photograph in which it appears with much distinction. The recorded weight of this gigantic

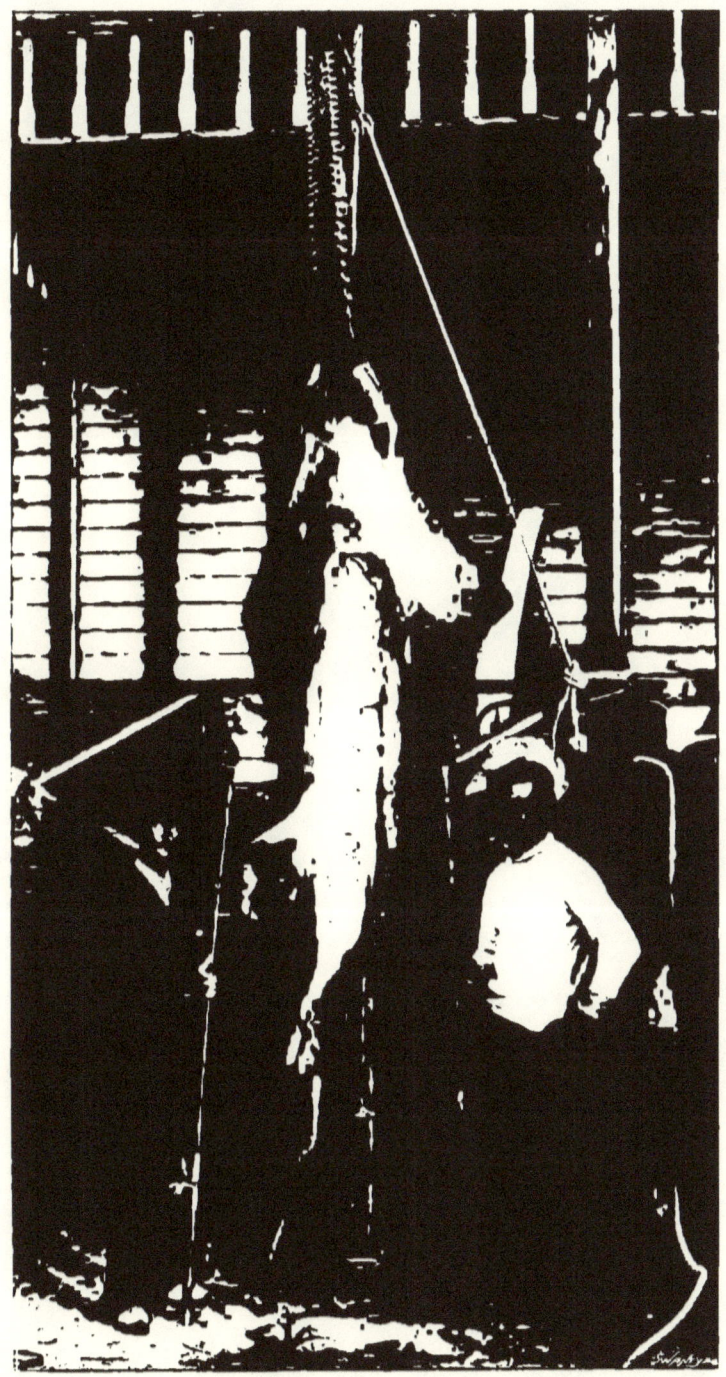

THE SAW-FISH AND ITS CAPTOR.

Length, 14 feet; breadth, 4½ feet. Estimated weight, 600 lbs. Florida, 2nd May 1897.

SAW-FISH TAKEN ON TARPON TACKLE.

THE SAME SAW-FISH GETTING ON TO SMALL BOAT.

marine perch was 796 lbs.   The kodak (see page 61) has per-
formed its duty admirably with the group, which, besides the Jew-
fish, consists of eight tarpon, the largest of which was 112 lbs.,
and the lady and gentleman who caught them the summer before
last at Captiva Pass.   Such monsters have to be hauled by
dead weight hand over hand from the bottom, which is generally of rocks, and no better testimony could be afforded of the
sterling qualities of the American-made linen lines.   This fish
is probably the black grouper.   There are two kinds, and both
are called " Jew-fish," and also " Warsaw."   Goode suggests
that these, which have been catalogued as *Epinephelus nigritus*
and *Promicrops guasa* respectively, are the young and adult of
the same species.

Of the former, Professor Baird wrote : " The black grouper
is resident all the year, though not abundant.   The greatest
size attained is about 15 pounds.   They pass the winter in
the salt-water rivers, living in holes in the rocks, and under
roots and snags, and about piles.   They are solitary in their
habits.   They feed on small fish, particularly mullet, and on
crustaceans, and breed in the salt rivers in May and June.
Their spawn is very small, and pale yellow.   They are taken
with hook and line by the use of mullet and crab bait, and are

seldom seen except when thus captured.   They are much
esteemed as food."

Of the latter, Brown Goode wrote [1] : " It is a fair question
whether this great fish be not the adult of the common black
grouper or some closely allied species, the appearance of which
has become somewhat changed with age.   A large specimen,
weighing about 300 pounds, was taken near the St. John's
bar in March or April 1874 by James Arnold.   It was
shipped by Mr. Hudson, a ship dealer in Savannah, to Mr.
Blackford, who presented it to the Smithsonian Institution.
A fine cast of this specimen graces the Fisheries Hall of the
National Museum.   Professor Poey, by whom the species was
named, states that in Cuba it attains to the weight of 600
pounds.   An old Connecticut fisherman, who was for many
years engaged in the Savannah Market fishery, states that
the Havana smacks often catch Jew-fish.   They are so voraci-
ous that when put into the well with the groupers they would
do much damage.   The fishermen have found it necessary,
therefore, to sew their jaws together before placing them with
other fish."

[1] *American Fishes*, p. 50.

SMALL FISHING

THERE have been several allusions in the foregoing pages to what is termed "small fishing." It really means the miscellaneous sport to be enjoyed by the angler with fishes otherwise than tarpon. Sharks come under the head of nuisances ; and such fish as the saw-fish and the monster marine perch may also be omitted from this category as exceptions to be treated independently. The following are amongst the species which the angler in Floridan waters may at any time find at the end of his line. Some of them are very queer fish indeed.

Mr. Von Blake, who was one of the enthusiastic anglers whose acquaintance we made, and who enjoyed that grand take **The Moon-** of Spanish mackerel with his fly-rod described on **Fish.** page 69, one day in April showed me half a dozen moon-fish (*Chætodipterus faber*) which had been caught

Channel Bass.

Squeteague.
Red Snapper.

Ravallia.

Sheepshead.

Crevalli.

Grouper.

"SOME OF THEM ARE VERY QUEER FISH."

in the fishermen's nets whilst procuring him mullet for bait. They are most curious-looking things, and I had great pleasure in preserving one of them.

This also on some portions of the coast is known as the horse-fish, and, as usual, is elsewhere given other names, such as sun-fish, and blunt-nosed shiner. The moon-fish are of no importance to the sportsman, but are an excellent article of food, and numbers of them ten or twelve inches in length are sent to the New York markets.

While at St. James's I saw for the first time a fish which bore some resemblance to our familiar English perch, having transverse bars and a spinous dorsal fin. **The Sheeps-** There were a variety of fish cruising around the   **head.** pier, and a young fellow was trying his best to harpoon some of them. We could only obtain a passing glimpse of these perch-like fish, and were not sure of their identity. My guide and captain, however, who were with me, pronounced that it must be the sheepshead (*Sargus ovis* or *Archosargus probato-cephalus*), one of the most valued food-fishes, and very likely it was, since a number of the sea-bream family are found along the shores of Florida throughout the entire year.

The ichthyological authorities say that in some of the

rivers of Florida the sheepshead becomes almost a fresh-water species, being constantly taken in seines in company with bass, perch, and suckers, far above the limits of perceptibly brackish water. It seems to derive its name from the somewhat sheepy expression imparted to its countenance by the peculiar mouth and teeth, and partly also from its innocent habits of browsing amongst subaqueous vegetation. Though not, strictly speaking, a sporting fish, it is to be reckoned amongst the genuine spoils of the small fisher. It is fished for mostly by the natives with some of the small crustacea—shrimps or crabs for choice if they are in season ; but oysters fresh from the shell, or even boiled, are found excellent bait.

This fish gave us plenty of pretty sport at the Captiva Pass and other fishing grounds. You might imagine from the **The Spotted** frequency with which the sacred name of trout is **Squeteague.** uttered in these fishing resorts of Florida that you are in Hampshire or Derbyshire. The reference, however, is invariably to the spotted squeteague, or weak fish, one variety of which is called the sea or deep-water trout in other parts of America. Our fish was the spotted squeteague (*Otolithus carolinensis* or *Cynoscion maculatus*), and very prettily spotted it was.

There are numerous varieties that are spoken and written of as trout or sea trout, but commercially the best known in America, where it regularly appears in the market, goes by the name of weak fish.

Another fish is the Crevalli or "cavally" (*Caranx hippos*), which is distributed throughout the West Indies, and is found on many coasts of the Pacific and Indian Oceans. **The** In Goode's work its sporting characteristics are **Crevalli.** very happily hit off in the following lines by Isaac McLellan, printed at the head of the chapter :

> Swift speed crevallé over that watery plain,
> Swift over Indian river's broad expanse.
> Swift where the ripples boil with finny hosts,
>   Bright glittering they glance ;
> And when the angler's spoon is over them cast,
> How fierce, how vigorous the fight for life,
> Now in the deeps they plunge, now leap in air,
>   Till ends the unequal strife.

The cavally belongs to the horse-mackerels. Another fish of the same family is the silver moon-fish (*Argyriosus vomer*).

I have never been in strange waters where I have not found a fish species called Jew-fish, and, roughly speaking, I

H

should say that in some parts of the world it is indifferently applied to every variety as to which nothing definite is known.

Jew-Fish. The Jew-fish of which I saw and photographed such a magnificent specimen is, as I have intimated, one of the groupers which occur on the reefs in Florida and West Indian waters. The specimens of a few pounds which we obtained in small fishing, seemed to me to be the black grouper before mentioned as *Epinephelus nigritus*.

The black bass (*Micropterus salmoides*) is too well known to need any description here, and I only mention it in passing to **Black Bass.** explain that it occurs in the inland waters and up the rivers of Florida, and will afford the usual sport to those who lay themselves out to angle for them. I caught quantities on several occasions with an artificial minnow whose triangles were loose. Somehow I repeatedly missed the fish which struck, but found that I caught every one after trying the experiment of tying the triangles close to the body of the bait.

The larger species of sea fish frequently caught in small **Channel Bass.** fishing is the channel bass or red drum (*Sciæna ocellata*), which is a much appreciated member of the Scianidæ, and claims much respect as being a strong, handsome fish, and own cousin to the maigre which we find in the Mediterranean and sometimes on the British coast. The Jew-fish of the Australian colonies is another relative. It is stated that " channel bass " is a characteristic name applied properly only to large individuals taken in channels, and that the smaller fish of the species are roughly called bass or school bass.   In Florida no distinction is made ; the fish is always called bass, just as the squeteague is always called trout.   They run in weight up to 50 lbs.

.The Ravallia is the snook of the Gulf coast.   It also bears the Spanish name of robalo, with a number of other designations. In many features it reminds one of the pike perch of Germany.   In science it is known as the **Ravallia.** *Centropomus undecimalis ;* its length is about 3 feet, and it is predatory to the backbone.   The ravallia ought to be a capital game fish, though Jordan says that it is seldom taken by the angler.   Snook seems rather an ignoble name to apply to a respectable raider amongst other fishes, but the word is a corruption of *snoek*, which is to this day the Dutch name for pike.

The snapper is, I think, correctly placed by Dr. Günther amongst the perches, though Jordan, the American scientist,

ranks them amongst the sea breams.   They are very abundant
around the American coast.   The red snapper, or grunt
(*Mesoprion aya*), as it is sometimes dubbed, is in everlasting

Red Snapper.   evidence in the Gulf of Mexico and Eastern
Florida, and has been for many years now highly
esteemed as an article of food in the northern markets to which
it is transported.   This fish is found with the sea bass, and its
range is as far north as Cape Cod ; but it is in the Gulf of
Mexico that the red snappers are most at home.   They haunt
the reefs and rocks, and are very plentiful off Charlotte
Harbour.   They give excellent sport, but are mostly taken
with the net for market.   I have heard of some excellent
takes of red snapper, which are carnivorous fish, with spinning
bait, and they sometimes run to 40 lbs. weight.   The specimens
which I saw, and occasionally caught, were, however, nothing
like so large ; the average is perhaps 8 lbs. and 9 lbs., and
a bit of dead fish is good bait.

A more gracefully shaped variety which often comes in
the way of the small-fish angler is the mangrove snapper
(*Mesoprion aurorubens*).   It is active and beautifully coloured
with russet and gold lines, and is seldom more than 18 inches
long.

BLACK BASS.

X

THE SEASON OF 1897

THE Diary which has been freely laid under contribution was continued to the end, but on the tolerably familiar ground of the train journey to Washington. **Homewards.** Even on our return we did not escape the casualties of American travel, for at one station we were compelled to wait four hours for another part of the train, which had been delayed owing to some breakdown, with the consequence that we arrived at Washington at eleven instead of 7.30.

After our experience on board *The Tarpon* at Fort Myers we were very glad to get back again to absolute civilisation, to look over the Natural History Museum and Smithsonian Museum, and to enjoy the avenue of trees, which in this early May began to look very charming. We saw plenty of typical American fish in the market next day, and on Sunday after church "did" the Zoo and Park. Next day we went over the Capitol, and heard some of the debates in the House of Representatives.

Next we took our camera and ourselves to Niagara, and, spite of rainy, cold weather, enjoyed the visit and took a quantity of admirable views. I noticed some people fishing at Lewistown where they catch pike, perch, and bass, especially at that time of the year, and again in the autumn.

On my return to England I received the *Fort Myers Press*, and learned from it that, after we left, a party of American

Sport in May. gentlemen arrived to continue the fishing. They commenced on 10th May, and caught tarpon at once. This bears out a previous remark of mine that, whether the fish are then in season or not, there is sport to be found, at the price of course of increasing heat.

Upon the question of spawning and other kindred subjects, I may quote an extract from Mr. C. Stewart Davison's contribution to *The Spirit of the Times* of 19th June 1897. In reviewing the season he says :—

The time necessary to kill a fish of course varies with their size, strength, and condition. A female fish shortly before spawning may contain as much as from 20 to 40 pounds of roe, be very sluggish and non-resistant, and be brought to gaff in but a few minutes. On the other hand, early in the season with a strong active fish, an hour, and even two and three hours have been necessarily employed, even by those fairly expert. It seems a pity to take tarpon, as any other fish, at or about the spawning time, but as they only visit the coast apparently for that purpose, it is difficult to see how it can be wholly avoided. All fish are, potentially, spawners as all animals are breeders, and increase of game is necessarily diminished by each success of the sportsman. Nevertheless, it would probably be desirable for future sport if a limit of time were fixed. Substantially the increasing heat of the weather in Southern Florida does fix a limit, as fishing practically ceases by the 15th of May, but probably 1st May would be better. The tarpon hardly falls within the class of edible fishes. They can, of course, be eaten, and the little ones, 40 pounds and under, are not as bad as the larger ones. But none are very good. This year at Fort Myers, out of 165 fish taken in all, to and including the 12th of May, 19 fish were under 70 pounds, the others running up to the largest of the season, 187 pounds taken on the 28th of April by Mr. W. Ashby Jones, of Richmond, Virginia. This fish measured 6 feet 11 inches in length, and it is very near the largest which has been taken on rod and reel. Mr. Jones's record, by the way, is the best for the season. Fishing (from 21st March to 11th May) on 45 days, he took 38 tarpon aggregating 4236 pounds, say 112 pounds per fish. Of these one was taken "trolling," *i.e.* on an exaggerated "Devon," trailed on the usual rod and reel behind a rowboat, the fish being hooked in the mouth and landed in seven minutes. Necessarily, this involves never allowing a fish to gain any slack, as hooking tarpon in the mouth is a very unsecure method, there being nothing in which the barb of the hook can set. This is better illustrated by the " Pass " fishing, which is wholly different from river fishing, but which, as this letter is reaching an undue length, I can hardly describe in detail.   It may be sufficient to say that "Pass" fishing is mostly practised, on the west coast, at Captiva Pass, a passage between two islands out to the Gulf.   The tide makes some miles an hour, and usually the breakers outside are heavy.   Your boat being anchored in the tide-way, some 30 yards of line are allowed to trail behind in the tide baited with either a gigantic spoon, enormous devon, or an entire mullet hooked through the mouth.   The tarpon running past strike and must be themselves struck instantly.   Immediately on striking a fish the boatman gets up anchor and rows for the beach (as otherwise boat, tarpon and all would go out through the Pass).   The fish must be landed on the beach, the gaffer going into the shallow water for this purpose, and from beginning to end as bait and hook have not been gorged, no slack line should be allowed the fish.   It has been suggested that "central draft" hooks would be the best for Pass fishing, and I believe such have been tried.

The *Fort Myers Press* publishes a record of the fish which **A Season's** are landed and weighed at that place, and its **Record.** substantial accuracy may, I think, be fairly taken for granted.   The Returns for the season are appended, and the reader will not fail to remark how uniformly heavy the fish were in the May month after my departure.

EXTRACT FROM *Fort Myers Press*

TARPON RECORD FOR 1897

The following is the record of tarpon caught at Fort Myers with rod and reel this season :—

	Pounds.	Ft.	In.
Jan. 11. A. T. G. Parkinson, England	1½	1	6
Feb. 12. J. W. Weber, Brooklyn, N.Y.	110	6	0
„ 13. B. Fallert, Brooklyn, N.Y.	135	6	6
„ 20. Jos. Fallert, Brooklyn, N.Y. .	100	5	9
„ 20. Jos. L. Hickson, Minneapolis, Minn.	91	5	10
„ 21. Geo. W. Straub, Brooklyn, N.Y.	87	5	6
Mar. 11. W. W. Olln, Kalamazoo, Mich.	123	6	0
„ 12. A. T. G. Parkinson, England .	2	1	7
„ 15. Jos. L. Hickson, Minneapolis, Minn.	96	6	4
„ 15. R. R. Rand, Minneapolis, Minn.	126	6	0
„ 17. J. J. Bartlett, Strongsville, Ohio .	112	6	0
„ 17. J. P. Wheelright, Minneapolis, Minn.	59	5	0
„ 17. M. B. Koon, Minneapolis, Minn. .	41	4	6
„ 19. M. B. Koon, Minneapolis, Minn. .	90	5	7
„ 19. S. H. Chisholm, Cleveland, O.	108	5	9
„ 19. F. B. Loebs, Rochester, N.Y. .	90	5	6
„ 21. Chas. Stewart Davison, New York City .	138	6	4
„ 21. Jos. L. Hickson, Minneapolis, Minn.	76	5	10
„ 21. Jos. L. Hickson, Minneapolis, Minn.	138	6	5
„ 21. R. T. Holloway, Lexington, Ky. .	70	...	
„ 21. W. Ashby Jones, Richmond, Va. .	150	6	4
„ 22. R. T. Holloway, Lexington, Ky.	134	...	
„ 22. R. T. Holloway, Lexington, Ky.	71	...	
„ 22. W. Ashby Jones, Richmond, Va. .	132	6	1

			Pounds.	Ft.	In.
Mar.	22.	Jas. H. Prince, Boston, Mass.	78	5	0
,,	23.	W. Ashby Jones, Richmond, Va.	134	6	4
,,	23.	W. Ashby Jones, Richmond, Va. .	125	6	2
,,	24.	Chas. Stewart Davison, New York City .	125	6	2
,,	24.	W. Ashby Jones, Richmond, Va.	99	5	6
,,	24.	Jas. H. Prince, Boston, Mass.	31	4	1
,,	12.	C. F. W. Millatz, New York .	55	4	0
,,	14.	C. F. W. Millatz, New York .	21	6	0
,,	21.	C. F. W. Millatz, New York .	90	6	0
,,	21.	Centre Hitchcock, New York	157	6	7
,,	23.	F. Gray Griswold, New York	151	6	9
,,	24.	O. A. Mygatt, New York .	150	6	9
,,	26.	W. Ashby Jones, Richmond, Va. .	81	5	8
,,	26.	W. Ashby Jones, Richmond, Va. .	90	5	10
,,	27.	Jos. L. Hickson, Minneapolis, Minn.	126	6	1
,,	27.	W. Ashby Jones, Richmond, Va. .	73	5	3
,,	27.	W. Ashby Jones, Richmond, Va. .	111	5	11
,,	29.	A. M. Mitchell, New York.	131	6	5
,,	29.	A. M. Mitchell, New York .	121	6	5
,,	28.	R. T. Holloway, Lexington, Ky. .	68	5	4
,,	31.	Miss Anna Holloway, Lexington, Ky.	90	5	6
April	3.	R. T. Holloway, Lexington, Ky. .	120	6	3
,,	3.	A. M. Mitchell, New York . .	140	6	7
,,	3.	Mrs. W. Ashby Jones, Richmond, Va.	97	6	0
,,	5.	Mrs. A. M. M'Gregor, New York .	94	6	1
,,	5.	A. M. Mitchell, New York	66	6	0
,,	5.	A. M. Mitchell, New York .	43	5	0
,,	5.	Jas. Loughlin, jr., Pittsburg, Pa.	150	6	4
,,	5.	R. T. Holloway, Lexington, Ky.	124	6	0
,,	5.	R. T. Holloway, Lexington, Ky.	100	5	11

			Pounds.	Ft.	In.
April	5.	W. Ashby Jones, Richmond, Va. .	. 102	6	4
,,	7.	W. Ashby Jones, Richmond, Va.	86	5	10
,,	7.	R. T. Holloway, Lexington, Ky.	126	6	0
,,	7.	R. T. Holloway, Lexington, Ky.	163	6	10
,,	7.	R. T. Holloway, Lexington, Ky.	74	5	4
,,	7.	James Loughlin, Pittsburg, Pa. .	75	5	4
,,	7.	Mrs. Frank Q. Brown, Boston, Mass.	75		...
,,	8.	Henry B. Plant, Plant System .	150	6	6
,,	9.	Jos. L. Hickson, Minneapolis, Minn.	93	5	8
,,	12.	R. T. Holloway, Lexington, Ky. .	114	6	0
,,	12.	Rowland Ward, London, England .	123	6	3
,,	13.	W. Ashby Jones, Richmond, Va.	90	5	1
,,	13.	W. Ashby Jones, Richmond, Va.	58	4	8
,,	13.	A. M. Mitchell, New York City	86	5	8
,,	13.	A. M. Mitchell, New York City	105	6	1
,,	14.	R. T. Holloway, Lexington, Ky.	119	6	2
,,	14.	W. Ashby Jones, Richmond, Va.	107	6	2
,,	14.	W. Ashby Jones, Richmond, Va.	124	6	0
,,	14.	A. M. M'Gregor, New York .	84	5	6
,,	15.	B. B. M'Gregor, New York .	105	6	2
,,	15.	B. B. M'Gregor, New York .	130	6	4
,,	16.	W. Ashby Jones, Richmond, Va. .	95	5	11
,,	16.	W. Ashby Jones, Richmond, Va. .	75	5	2
,,	16.	Mrs. R. J. Shepherd, Fort Myers, Fla.	116	6	0
,,	16.	R. T. Holloway, Lexington, Ky.	136	6	3
,,	16.	R. T. Holloway, Lexington, Ky.	97	6	1
,,	16.	F. S. Hodges . . .	143	6	6
,,	16.	R. T. Holloway, Lexington, Ky.	. 94	5	0
,,	17.	W. Ashby Jones, Richmond, Va. .	110	5	10
,,	17.	W. Ashby Jones, Richmond, Va. .	81	5	3

		Pounds.	Ft.	In.
April 19.	R. T. Holloway, Lexington, Ky.	135	6	6
„ 20.	R. T. Holloway, Lexington, Ky.	115	5	11
„ 20.	R. T. Holloway, Lexington, Ky.	160	6	8
„ 17.	John T. Spaulding, Boston .	102	5	10½
„ 19.	John T. Spaulding, Boston	160	7	1½
„ 15.	George Mixter, Boston .	162		
„ 17.	George Mixter, Boston .	30		...
„ 20.	Rowland Ward, London	114	6	0
„ 13.	O. A. Mygatt, New York	60		
„ 13.	O. F. W. Millatz, New York .	100		
„ 14.	O. F. W. Millatz, New York .	135		
„ 17.	O. A. Mygatt, New York	121		
„ 17.	O. A. Mygatt, New York .	102		...
„ 18.	W. Ashby Jones, Richmond, Va.	117	6	4
„ 20.	O. F. W. Millatz, New York .	140		
„ 20.	O. F. W. Millatz, New York .	60		...
„ 21.	W. Ashby Jones, Richmond, Va.	64	4	12
„ 23.	Rowland Ward, London, Eng.	44	4	6
„ 24.	Rowland Ward, London, Eng.	92	5	8
„ 24.	W. Ashby Jones, Richmond, Va.	140	6	7
„ 26.	W. Ashby Jones, Richmond, Va.	187	6	11
„ 26.	R. T. Holloway, Lexington, Ky. .	35	4	6
„ 27.	P. J. M. Van Cortlandt, New York	132	6	1
„ 27.	R. T. Holloway, Lexington, Ky.	120	5	11
„ 27.	R. T. Holloway, Lexington, Ky.	61	5	0
„ 27.	R. T. Holloway, Lexington, Ky.	87	5	8
„ 27.	R. T. Holloway, Lexington, Ky. .	168	6	3
„ 28.	P. J. M. Van Cortlandt, New York	108	5	1½
„ 28.	P. J. M. Van Cortlandt, New York	103	5	8
„ 28.	R. T. Holloway, Lexington, Ky.	156	6	5

	Pounds.	Ft.	In.
April 28. R. T. Holloway, Lexington, Ky. .	148	6	5
„ 28. Rowland Ward, London, Eng. .	122	6	0
„ 27. W. Ashby Jones, Richmond, Va. .	172	6	9
„ 26. A. C. Craig, New Jersey .	131	6	3
„ 27. A. C. Craig, New Jersey	111	6	0
„ 27. A. C. Craig, New Jersey	81	5	11
„ 27. O. A. Mygatt, New York . .	125	...	
„ 29. W. Ashby Jones, Richmond, Va. .	154	6	9
May 2. W. Ashby Jones, Richmond, Va.	149	6	9
„ 2. W. Ashby Jones, Richmond, Va.	123	6	1
„ 2. W. Ashby Jones, Richmond, Va.	134	6	5
„ 3. W. Ashby Jones, Richmond, Va.	126	6	1
„ 3. W. Ashby Jones, Richmond, Va.	103	6	0
„ 2. Geo. L. Shipley, Providence, R.I. . .	123	6	2
„ 2. Edward Vom Hofe, New York (saw-fish)	600	14	0
„ 4. Edward Vom Hofe, New York .	135	6	1
„ 4. P. J. M. Van Cortlandt, New York .	156	6	6
„ 3. Edward Vom Hofe, New York (saw-fish)	500	13	0
„ 4. Thos. E. Peters, London, England .	70	5	4
„ 4. W. Ashby Jones, Richmond, Va. .	100	6	0
„ 5. P. J. M. Van Cortlandt, New York	131	5	9
„ 5. E. Vom Hofe, New York .	100	5	10
„ 5. W. Ashby Jones, Richmond, Va. .	107	5	10
„ 5. W. Ashby Jones, Richmond, Va. .	130	6	4
„ 5. W. Ashby Jones, Richmond, Va. .	144	6	10
„ 6. P. J. M. Van Cortlandt, New York	57	4	6
„ 6. E. Vom Hofe, New York . .	122	6	1
„ 6. W. Ashby Jones, Richmond, Va. .	74	5	2
„ 7. Judge Emmett Field, Louisville, Ky.	125	6	0
„ 7. Mrs. M. L. Satterwhite, Louisville, Ky. .	107	5	11

			Pounds.	Ft.	In.
May	7.	E. Vom Hofe, New York .	134	...	
„	8.	E. Vom Hofe, New York . .	145	...	
„	8.	Mrs. M. L. Satterwhite, Louisville, Ky. .	142	6	6
„	8.	Mrs. Bennett H. Young, Louisville, Ky..	110	5	9
„	8.	W. Ashby Jones, Richmond, Va. . . .	79	5	9
„	8.	W. Ashby Jones, jun., Richmond, Va. (boy 12 years old)	98	5	11
„	8.	Judge Emmett Field, Louisville, Ky.	80	5	8
„	9.	W. Ashby Jones, Richmond, Va.	111	6	5
„	10.	W. Ashby Jones, Richmond, Va.	99	6	0
„	10.	Henry Kaufman, Louisville, Ky. .	158	6	5
„	11.	Col. Bennet H. Young, Louisville, Ky.	163	7	3
„	11.	W. T. Grant, Louisville, Ky.. .	107	6	0
„	11.	G. W. Lowman, Louisville, Ky.	164	6	11
„	11.	W. H. Nevins, Birmingham, Ala. .	139	6	4
„	11.	W. H. Nevins, Birmingham, Ala. .	124	6	2
„	11.	W. H. Nevins, Birmingham, Ala. .	79	5	6
„	11.	A. H. Carpenter, Birmingham, Ala. .	119	...	
„	12.	Major Frank Y. Anderson, Birmingham, Ala. .	60	5	6
„	12.	Daniel H. Rogan, Birmingham, Ala.	131	6	7
„	12.	Daniel H. Rogan, Birmingham, Ala.	106	6	0
„	12.	Wm. H. Martin, Birmingham, Ala.	93	5	8
„	12.	A. H. Carpenter, Birmingham, Ala. .	105	6	0
„	13.	Major Frank Y. Anderson, Birmingham, Ala. .	95	6	1
„	13.	Daniel H. Rogan, Birmingham, Ala.	69	5	7
„	13.	Wm. H. Kettig, Birmingham, Ala.. .	106	6	0
„	14.	Major Frank Y. Anderson, Birmingham, Ala. .	135	6	8
„	14.	Major Frank Y. Anderson, Birmingham, Ala. .	127	6	7
„	14.	Major Frank Y. Anderson, Birmingham, Ala. .	69	5	6
„	14.	Major Frank Y. Anderson, Birmingham, Ala. .	65	5	1

			Pounds.	Ft.	In.
May	14.	Jack W. Johnson, Birmingham, Ala.	120	6	1
.,	14.	Jack W. Johnson, Birmingham, Ala.	151	6	6
,,	14.	Chas. W. Ferguson, Birmingham, Ala.	123	6	6
,,	14.	Geo. R. Harris, Birmingham, Ala. . '	108	6	8
.,	15.	E. W. Wilkinson, Birmingham, Ala.	135	6	6
,,	15.	E. W. Wilkinson, Birmingham, Ala.	63	5	4
,,	7.	O. A. Mygatt, New York . .	118		
,,	8.	O. A. Mygatt, New York	60		
,,	10.	O. A. Mygatt, New York	110		
,,	12.	O. A. Mygatt, New York .	128		
,.	10.	O. F. W. Millatz, New York .	100		
,,	12.	O. F. W. Millatz, New York ·	135		

There appear to be no records from Punta Rassa or Naples, but the following are returns of early fishing at Marco and St. James' City. **Marco and St. James.**

TARPON RECORD AT MARCO

			Pounds.	Ft.	In.
Feb.	18.	Hugh Macdonald, jun., Covington, Ky. .	97	5	8
Mar.	3.	L. Bury, England	123	6	3
,,	5.	Hugh Macdonald, jun., Covington, Ky. .	110	6	0
,,	7.	Hugh Macdonald, jun., Covington, Ky. .	116	6	0
,,	7.	Frank M. Hoyt, Milwaukee, Wis. .	97	5	10
,,	8.	Hugh Macdonald, jun., Covington, Ky.	116	5	9
,,	8.	Hugh Macdonald, jun., Covington, Ky. .	90	5	8½
,,	8.	Henry W. Vaughan, New York .	118	6	4
,,	9.	A. J. Cunningham, Rochester, N.Y.	130	6	4
,,	11.	J. A. Jameson, London, England .	135	6	5

packing case, preferring it to the heavy, hard, expensive wooden affairs like a coffin, stuffed with very heavy material, for which you have to pay high rates to the railway company.   My box cost nothing as extra baggage, but came as personal luggage.   The great point in skinning a tarpon for future mounting is to have the grease removed from the inside, otherwise a beautiful silver fish becomes quite black in time.

OTHER SPORT IN FLORIDA

AMONGST the works which will assist the sportsman who meditates a visit to Florida is *Hunting and Fishing in Florida, including a Key to the Water Birds known to occur in the State*, by Chas. B. Cory (Estes and Lauriat, Boston, Massachusetts, 1896). Mr. Cory is a well-known ornithologist, and has been hunting, fishing, and shooting in Florida since 1877, while each of the ten winters he spent in the State was devoted to exploring out-of-the-way nooks and corners, which ten years ago were known to very few people. Although Florida is now visited annually by thousands instead of hundreds of people, as was the case a few years ago, there is still a vast extent of country which is practically a wilderness, and where game is yet to be found in great abundance.

The following summary, for some of which I am much

indebted to Dr. Bean, of the Museum at Washington, Mr. A. N. Cheney, the Fish Commissioner for New York

A QUIET SPOT.

State, as well as to the work to which reference has just been made, may be useful to readers who desire an outline of knowledge as to the beasts, birds, and reptiles of Florida :—

The Seminole Indians, whose manners and customs have been closely studied and described by other authors, are fond of hunting the Manatee (*Manatis americanus*), one **The** of the sea-cow family represented by the dugong **Manatee.** in Australia. These creatures enter the rivers to browse on a special kind of marine grass, and some writers say also on the leaves of the mangroves. The manatee is abundant in the bays and rivers all along the east and west coast of southern Florida. I heard of them, but never had the opportunity of seeing one during our wanderings. They are said to live equally in salt or fresh water, and the Indians kill numbers every year, hunting them in canoes generally near the mouth of some river, and harpooning them as they rise to the surface.

The Florida puma (*Felis concolor*), still not uncommon in the unsettled parts, is somewhat smaller and more rufous in colour than its northern and southern brethren, **The** and though very shy and wandering in its habits, **Puma.** preys upon small animals, and even deer and dogs. In one week, within thirty miles of Lake Worth, a hunting party

I 2

found the fresh trails of seven animals. So lately as the winter of 1895, pumas were quite numerous near the Cypress Swamps, about Long Hammock, Custard Apple Hammock, and the parts south-west of Lake Worth. A puma 9 feet long may be considered a big one in Florida. Estimated weight about 130 lbs.

The Black Bear (*Ursus americanus*) is still found in some parts of Florida, especially near the coast. Old hunters **The** speak of specimens up to 600 lbs. Good sport **Black Bear.** may be had hunting bears in Florida, but to do it successfully an efficient pack of hounds is required, and one or two of them must be thoroughly trained bear dogs. The bears hibernate in Florida from the end of December until March, and then they are found hunting for crabs along the shore, eating the young palmetto cabbage, and the seeds (or " buds ") of the mangroves. In June they haunt the beaches for turtles' eggs, and in the fall they feed upon the palmetto berries, which grow in abundance on the sandy lands bordering the ocean. One specimen, shot through the lungs with a bullet from a 45-70 Winchester by Mr. Cory in December 1893, was an old male and very fat, weighing about 500 lbs. (estimated) ; it measured 6 ft. 2 in. from nose to tail in a straight line, 8 ft. 6 in. from hind claw to nose, and 54 inches around the chest; and one of the front claws was $3\frac{1}{4}$ inches, measured on the curve. This is described as a fair example of the Florida bear.

The haunts of the Florida deer (*Cervus virginianus*) are becoming fewer every year. They are still plentiful in remote parts, but it is rare now to see eight or ten feeding on a prairie at one time. They are **Deer.** hunted principally on horseback with hounds; by slow trailing; and a third method is still hunting, after the manner of the Indians. The Florida variety is smaller and of a slightly different colour from the true Virginian deer; and a full-grown buck will not often weigh over 110 lbs., and stands about 24 inches at shoulder. The young are born in April and May.

The Florida wild cat, a variety of the northern form (*Felis rufa*) is fairly numerous, and some specimens **Lynx or** stand 18 or 20 inches at the shoulder. **Wild Cat.**

The wolf (*Canis lupus*) is still found in restricted localities, especially in the vicinity of the Big Cypress Swamp, and in extreme Southern Florida. They **Wolves.** are, I am told, getting exceedingly scarce, and specimens of the black variety are very handsome skins.

Amongst other animals are the gray fox (*Canis virginianus*), the weasels (*Mustelidæ*), the mink (*M. visor*), the otter (*Lutra*

Other small *canadensis*), the common skunk (*Mephilitis mephi-*
Mammals. *tica*), the little striped skunk (*Spilogaleputorius*), the only climbing example of this group—the raccoon (*Procyon lotor*), bats of various kinds, shrews, moles, the marsh rabbit, the gray rabbit, gopher or salamander, a variety of rats, mice, flying and other squirrels, and opossum (*Didelphys marsupialis*).

Alligators have been previously mentioned as objects seen by us during our stay in Florida. Skin hunters kill their

alligators at night, using a light, which enables
Alligators. the boat to approach within a few feet of the animal, which is then shot without difficulty. When the alligators have not been disturbed, and have become tame and lazy, they may be approached by quietly paddling up the creeks and rivers, and by sending a shot into the eye or any portion of the head, so as to penetrate the brain; occasionally they may be found far inland. Anything over 12 feet is a good one.

Crocodiles occur in the rivers and bays of extreme Southern Florida, but are rarely found very far from salt water. They grow to a larger size than the alligator. The largest crocodile Mr. Cory mentions was 13½ feet in length.   **Crocodiles.**

There are four snakes classed as venomous : they are locally known as the diamond rattlesnake (*Crotalus adamanteus*), average length about 6 feet ; the ground rattle-snake (*Sistrurus miliarius*), average length about   **Snakes.**
3 feet ; the stump-tailed moccasin (*Agkistrodon piscivorus*), venomous but uncommon ; and the harlequin snake (*Elaps fulvius*), a small species distinctly poisonous. The non-poisonous snakes number between twenty-five and thirty, and include the water moccasin, the thunder snake, the king snake, the black snake, the coach whip, the chicken snake, gopher snake, hognose snake, and garter snake.

My sport in Florida was principally confined to the rod, the season for the gun being during the winter months. Making all allowance for the decidedly tall stories   **Game and** one hears and reads of bags of so-called turkeys, **other Birds.** quail, wild ducks, and snipe during the season between fall and spring, I have no doubt the fowler would find an abundance of good shooting. The Darters, or so-called water turkeys (*Plotus anhinga*), are very numerous in some parts, but have been gradually exterminated in certain localities by native hunters who trap them.

Quail (*Ortyx virginianus*) and snipe (*Scolopax*) may be found within easy driving distance of many of the pleasure

Quail.

resorts, and are abundant throughout the state, especially in the southern portion. Dogs are, of course, necessary.

Migratory wild fowl of all kinds once visited Florida in incredible numbers, and over decoys or stools a bag of one hundred birds in a day's shooting was not uncommon. The birds are now scarcer, but the regular visitors still include wigeon (*Mareca americana*), pintail (*Dafila acuta*), shoveller (*Spatula clypeata*), blue and green wing teal (*Nettion carolinensis* and *N. discors*), ringneck (*Aythya collaris*), ruddy duck (*Erismatura rubida*), black Scoter (*Œdemia americana*), and blue bill or greater scaup.

Other birds one frequently sees are several varieties of bitterns principally (*Botaurus lentiginosus*), egrets (*Ardea candi-*

Miscellaneous Birds.

dissima), the white, blue, Ward's and Würdemann's herons, ibis, cranes, rails, sandpipers, curlews, avocets, golden and other plovers, oyster catchers, surf-birds, turnstones, and great northern divers.

REPRINT FROM *The Field*

30*th July* 1898

BOCA GRANDE'S GREAT TARPON FISHING

Sir — You noticed in your issue of 16th July Mr. and Mrs.
Holloway's tarpon record of this season, fishing from 15th March
till 15th May, 101 fish, largest 177 lb. ; but what do you think of two
rods leaving New York 15th May, returning again in June to New York
with a record of 120 tarpon—one rod apiece? Mr. O. A. Mygatt, who
is well known out there, and whom I have often seen with my glasses
working hard above Fort Myers from morn till night, either hunting or
fishing for tarpon, was one of them. On 18th May he and his friend
commenced fishing at Captiva Pass, but were not satisfied and went to
Boca Grande on the north side, trolling with the white part of a
mullet's belly. They saw hundreds of tarpon chasing fish, and playing,
but the sharks were very numerous and robbed them of many fish after
playing them for some time. Boca Grande is eight miles from Captiva
Pass. The water is very deep here, sixty feet, with hot tides, but there
is good anchorage inside for big vessels. Fishing one day, a mile
outside, and later in the day inside, Mr. Mygatt caught nineteen tarpon
and his friend ten, losing three others by sharks. He says it never
entered into his mind that he would have a better day than this. The
next day they were so tired out they rested. Four days later, in ten
hours, they killed thirteen tarpon, two measuring respectively 6 ft. 10 in.
by 42 in. girth and 6 ft. 11½ in. by 44 in. girth ; the next day Mr. O. A.
Mygatt got twenty-one tarpon to his own rod.

Every fish, he says, was played squarely to the finish, without any aid whatever.   Only three tarpon out of the twenty-one were gaffed. He says he found a new way to land them, catching hold of the wire snell and bringing their heads up on the sand, when it is possible to take the hook out and let the fish go back into the water.   It was this part of the sport that, when in Florida, I liked the least, viz. having to kill every fish and throw it into the gulf; but how Mr. Mygatt did it is easier to say than do.   He also speaks of another way, from the boat : as soon as the tarpon is tired and lying near the boat on its side, you play him hard for two or three minutes, and hold his head at every chance flush till he is played out.   The guide can then take the hook out, and let the fish go.   This must mean a terrific lot of force, I am sure.   The guide, in this case, wears a big thick glove on his left hand.

One day the anglers killed thirty-seven for the two rods; the physical endurance must be very great, as the heat is overpowering.   Previous to this Mr. Mygatt's best day's fishing, in ten hours, was eight fish.   The jump to nineteen and twenty-two proves that the Boca Grande is the best place to go, and my reason for writing you is that it has only just been found out.   It may interest English owners of yachts who will go there, to know that vessels drawing up to 20 ft. of water can get into the Pass and find good anchorage inside, where there are several square miles of water.   Mr. Mygatt, I can say from my own experience, spends a lot of time in hunting for tarpon long before he attempts to fish, and that is the way to go about tarpon fishing.   They show themselves if you will search for them.   Mr. Mygatt says that strong and stiff rods, of 6 ft. 8 in. to 7 ft., are best for killing your fish there.

Any one reading this must not forget when these bags were made, viz. 15th May to 18th June, including travelling from New York.   Later it is hotter, and earlier in the season the fish are not so numerous.

Many people go in January, but that is a mistake. As I have fully explained in my recent book, *The English Angler in Florida*, there is a lot of other fishing to be found in this place, of the kind which the Americans call "small fishing."   ROWLAND WARD.

166 PICCADILLY, W.,
   25th July.

[Mr. Rowland Ward's book, being an account of his own fishing in Florida last year, is literally "up-to-date," and the above letter is therefore a supplementary record, bringing us to what is practically the end of this year's tarpon season.—ED.]

www.ingramcontent.com/pod-product-compliance
Lightning Source LLC
Chambersburg PA
CBHW032009010726
47493CB00007B/2341